BEAM

JEFF HECHT

BEAM

THE RACE TO MAKE THE LASER

OXFORD
UNIVERSITY PRESS

2005

OXFORD
UNIVERSITY PRESS

Oxford University Press, Inc., publishes works that further
Oxford University's objective of excellence
in research, scholarship, and education.

Oxford   New York
Auckland   Cape Town   Dar es Salaam   Hong Kong   Karachi
Kuala Lumpur   Madrid   Melbourne   Mexico City   Nairobi
New Delhi   Shanghai   Taipei   Toronto

With offices in
Argentina   Austria   Brazil   Chile   Czech Republic   France   Greece
Guatemala   Hungary   Italy   Japan   Poland   Portugal   Singapore
South Korea   Switzerland   Thailand   Turkey   Ukraine   Vietnam

Copyright © 2005 by Jeff Hecht

Published by Oxford University Press, Inc.
198 Madison Avenue, New York, New York 10016
www.oup.com

Library of Congress Cataloging-in-Publication Data
Hecht, Jeff.
Beam : the race to make the laser
/ Jeff Hecht.
p.   cm.
Includes bibliographical references and index.
ISBN-13 978-0-19-514210-5
ISBN 0-19-514210-1
1. Lasers—History—20th century.   I. Title.
TA1677.H42 2004
621.36'6'09—dc22   2004002694

1   3   5   7   9   8   6   4   2

Printed in the United States of America
on acid-free paper

PREFACE

THE LASER RANKS with the transistor and the computer as a symbol of modern technology. The laser beam was something new to the world; light that was neatly ordered. A laser beam is a drill team in identical neatly pressed uniforms marching at halftime straight across the field in formation, not the disorderly mob of people streaming from the stadium afterwards. A laser beam is coherent light, waves of the same size, with their peaks and valleys lined up, following the same single path so straight and narrow that we call it "laser sharp."

This book tells the story of that remarkable invention. It focuses on an intense race to build the laser that took three years. The race began with a conversation between two very bright men who each dreamed about a source of coherent light they had not yet named. Charles Townes was an eminent physicist who had already invented the microwave counterpart of the laser. Gordon Gould was an easy-to-underestimate long-time graduate student who dreamed of being an inventor. The two went their separate ways, making waves that spread across the world of physics. Others joined the race, chasing a Nobel prize or a pot of gold at the finish line, but when the dust settled a dark horse named Theodore Maiman had crossed the finish line far ahead of the pack. Maiman's elegant design surprised the rest, leaving the prestigious Bell Laboratories in stunned denial of his success. It should have been over then, but Maiman had to battle for acceptance as the man who made the first laser.

Others have written ably about the history of the laser (see Sources). Maiman and Townes have written autobiographies, and Nick Taylor has chronicled Gould's success in a decades-long patent struggle. Joan Bromberg and Mario Bertolotti wrote scholarly histories of laser development. I have read and can recommend them all. But I have chosen a narrower focus: on the three dramatic years from the time Townes and Gould launched the high-stakes race for the laser to the recognition of Maiman as the winner. It's a story of science, invention, ambition, and scientists, set in an uneasy time, and destined to change their own lives and the world around them.

I owe profound thanks to many people who helped me in many different ways. Tony Siegman and Charlie Asawa have patiently answered questions, served as a sounding board for my ideas, read my draft manuscript, and generally steered me away from follies I was about to commit. Bill Bennett, Steve Jacobs, Paul Rabinowitz, Irwin Wieder, Irnee D'Haenens, Carol Botteron, Emile Tobenfield and Lois Hecht have read and commented on parts or all of my draft manuscript, and patiently responded to my questions.

A long list of people have given generously of their time to share their recollections with me in person or on the phone. Others have corresponded with me electronically from around the world, thanks to the Internet's ability to make links with people whom I otherwise could never find.

The late Bela Lengyel gave me valuable counsel in the early stages of my research.

The Sources section lists the name of the many people who gave generously of their time and energy to share their recollections with me in person, on the phone, in writing, or via electronic mail. This book could not have come alive without them. I owe them great thanks.

A few people deserve thanks for special bits of help. Adrian Popa gave me a grand tour that brought Hughes Research Laboratories to life. Gennady Gorelick helped me to understand the life of scientists in the Soviet Union. David Robertson, John Walko, Malcolm Butler, and Bob Cowan helped track down an explanation for the mysterious publication of Ted Maiman's laser paper in *British Communications & Electronics*. Nick Taylor generously shared information he had gathered for his excellent book *Laser: The Inventor, the Nobel Laureate, and the 30-Year Patent War*. Howard Rausch long ago brought me into the fascinating world of the laser community.

My research was greatly helped by access to the rich collection of documents and photographs held by the Center for the History of Physics at the American Institute of Physics. Their archives of interviews and documents collected by Joan Bromberg and Paul Forman for the Laser History Project are a wonderful time capsule of insight into the early years of the laser. The University of California's on-line oral history interviews with Charles Townes and Arthur Schawlow are a great resource. Ed Eckert excavated helpful documents and photos from the Bell Labs archives.

A generous grant from the Alfred P. Sloan Foundation as part of its program to encourage books on science and technology allowed me to set aside the time to write this book and do the detailed research necessary to make it credible. My agent Jeanne K. Hanson, and Kirk Jensen, Cliff Mills, John Rauschenberg, and Lisa Stallings of Oxford University Press, and Doron Weber of Sloan also deserve thanks for their help.

Memories fade with time, so I have tried to correlate documents with memories whenever possible. The mistakes that doubtless remain are my own.

CONTENTS

BEAM

PROLOGUE:
MAY 16, 1960, MALIBU, CALIFORNIA

IT WAS A MOMENT OF TRUTH for Ted Maiman. After months of careful analysis, he was ready to test his theory that a crystal of synthetic ruby could generate a new form of light, laser light, with its waves marching in phase like a drill team in formation. He and his assistant Irnee D'Haenens had set up their experiment in a laboratory the size of a two-car garage at Hughes Research Laboratories. Across the hall of the new building were offices with a spectacular view of the Pacific, but their windowless laboratory hid them from the outside world. An inevitable tension gnawed at Maiman. He had checked and double-checked everything, but no matter how carefully a scientist plans, nothing can guarantee that an experiment will succeed.

The physics establishment was convinced that Maiman's experiment was doomed. At first, Maiman himself had believed the eminent authorities who had declared that a ruby laser couldn't work. Yet a closer look convinced him the authorities were wrong. Maiman carefully planned a novel design that was both heretical and disarmingly simple. Instead of a bulky tube filled with rare gases or exotic metal vapors, he had built a device small enough to hold in the palm of one hand. Instead of trying to generate a feeble continuous signal at invisible infrared wavelengths, he aimed for bright pulses of red light.

The risk was undeniable. Yet so was the reality that no one else had succeeded in building a laser. It was not for lack of effort. The prestigious Bell Telephone Laboratories was working on several approaches. So was a classified program with a generous Pentagon contract. Both had million-dollar budgets, far beyond Maiman's. Some specialists had begun to suspect that lasers might be impossible to build. Maiman could be a bit headstrong, but facing the challenge required self-confidence and determination.

A fingertip-sized ruby rod sat inside the coil of a spring-shaped photographer's flashlamp, which itself sat inside a metal cylinder. Maiman hoped flashes from the lamp would excite the ruby to emit red pulses. A light-sensitive tube stood poised to convert light that hit it into electrical pulses that Maiman could display on the screen of a special oscilloscope. It wasn't enough to generate some light; ruby normally glows red when illuminated by bright light. Maiman believed that he could build the dim red glow into bright pulses by making the energized ruby amplify light that bounced back and forth between reflective silver films on the ends of the little rod. He wanted to push the little crystal beyond the threshold, where its normal red fluorescence became the first flash of laser light.

The idea of the experiment was to see what happened as they turned up the voltage across the lamp. The higher the voltage, the brighter the flash it should emit. They started by firing about 500 volts into the lamp. Then they turned it up one notch, then another. Each time they cranked up the voltage, the lamp fired brighter, and the ruby rod emitted more red fluorescence. A bright line on the oscilloscope screen traced the rise and fall of each red pulse, revealing the nature of the light. At first, it was the ordinary fluorescence they had expected. As they turned up the power, the fluorescence grew brighter. White light from the lamp leaked out from the little cylinder. The combined flashes dazzled the men's eyes, leaving them partly flash-blinded.

Maiman watched the screen intently, knowing that the oscilloscope traces would be the judge of success. The vertical rise of power on the screen was important, but the real test for laser action was the shape of the pulse shown on the scope. If the ruby crystal started firing laser pulses, the power level should rise and fall much faster than mere fluorescence. When they turned the voltage above 950 volts, the oscilloscope screen changed dramatically. "The output trace started to shoot up in peak intensity and the initial decay time rapidly decreased. Voilá. This was it! The laser was born!" Maiman wrote in his autobiography.

Theodore Maiman and Irnee D'Haenens display the first laser a quarter-century after they made it. (Hughes Research Laboratory photo, Courtesy AIP, Emilio Segre Visual Archives)

The color-blind D'Haenens was the first to see the horseshoe-shaped red glow of the laser light hitting the white cardboard screen. With few red sensors in his eyes, he could not see the faint red fluorescence. The flashes below laser threshold were bright enough to dim Maiman's normal color vision, but the laser's red pulses were so bright that even D'Haenens's color-blind eyes could respond. It was symbolic of the whole new type of light the laser represented.

D'Haenens jumped for joy; in his mid-20s, he was exuberant with the success. Maiman, more reserved and intense at 32, recalls he "was numb and emotionally drained from all the tension and excitement." He had weighed the odds, and had bet heavily. Now he had won the race to make the first laser.

It was an achievement to be proud of. Maiman had not merely beaten others trying to translate a well-defined theoretical proposal into reality. He had devised a new and different approach and made it work. He had invented a new type of laser, and made it work before anyone succeeded in demonstrating the original version.

The establishment was startled to see a dark horse standing on the far side of the finish line. They had entered some thoroughbreds in the race. Charles Townes had earlier invented the laser's microwave cousin, the maser, and had coauthored a detailed analysis of how what he called an "optical maser" should work. The maverick inventor Gordon Gould had devised his own plans, and snared a million dollars from the Pentagon for a crash program to build a laser. The prestigious Bell Labs had four groups of top physicists working on separate laser projects. None of them had expected Maiman to win the race.

This is the story of that race, and the people who ran it.

1

THE LASER RACE

THE RACE THAT TED MAIMAN WON to make the laser was neither simple nor easy. The starting line for the laser race was not obvious, and the course was undefined. Even the finish line—the world's first laser beam—was not clearly defined. Such is the quest for knowledge. You don't know where you're going until you get there, and if the goal looks interesting, you may have a lot of company heading in the same general direction.

Maiman faced tough competition. Bell Telephone Laboratories was the world's premier industrial research organization, with deep pockets and an impressive roster of scientific talent. The Pentagon had even deeper pockets and a wide range of military contractors to tap. Columbia University had a Nobel-studded physics department that could hardly be matched anywhere in the world. Westinghouse Electric, IBM, United Aircraft, American Optical, and other large American corporations also were working on lasers. Something was going on behind the Iron Curtain, although no one in the West was sure what.

Word of Maiman's success stunned the physics establishment. He was a little-known outsider working at a southern California aerospace company owned by the fabulously rich and eccentric Howard Hughes. He used ruby, a material which couldn't work in a laser, according to one of the young field's early experts, Arthur Schawlow of Bell Labs. Maiman broke the news at a Hughes press conference, not at a top scientific meeting or in a research journal. Some competitors initially couldn't believe Maiman had succeeded.

Yet Maiman's laser quickly passed the acid test of science—it could be reproduced. Within weeks, other physicists demonstrated their own ruby lasers. They didn't need detailed instructions, just press clippings and copies of a Hughes publicity photo that showed a different flashlamp. That rapid reproduction was an impressive confirmation of Maiman's simple and elegant design. It put the laser on the fast track to success.

Sadly, fate conspired to steal some of the credit Maiman deserved. The editors of the prestigious *Physical Review Letters* summarily rejected his report of the first laser, in one of the worst publishing blunders in the history of physics. Then a British engineering magazine published Maiman's paper without authorization, blocking its publication elsewhere. Bell Labs, adept at gaming the system, got a carefully worded report of experiments with a ruby laser—based on Maiman's design—into *Physical Review Letters*, the first report of a laser in an American journal. Later, Bell rubbed salt into the wound by aggressively claiming to have "invented" the laser.

It might smell like a conspiracy, but in reality it was a mixture of bad luck and aggressive self-promotion. Bell Labs employed a top-notch public relations department to polish the corporate image, and justify the parent AT&T's large research budget. Bell didn't tell out-and-out lies, but they did engage in typical puffery, inflating their claims and downplaying the work of others. Bell scientists made important contributions to laser research, but so did Maiman and others.

Theodore Maiman unquestionably made the first laser. He also invented the ruby laser, the particular type of laser that worked first. But no single person or group invented the laser, at least in the broad sense of an "invention" as a single master plan sufficient to make any laser work. Schawlow and Charles Townes outlined the basic principles of the laser while Townes was consulting at Bell Labs. Yet Gordon Gould outlined an even broader range of laser principles at the same time—as he was writing a patent application after leaving Columbia University. And neither proposal contained a complete blueprint to build a laser, although both described many concepts essential for laser operation. Both told how to arrange a pair of mirrors to make a laser, but the mirrors alone were not enough.

The laser was not a simple invention. It was the discovery of how to control the interaction of energy and matter to produce a beam of well-controlled light. It wasn't developed for any specific purpose. Bell Labs thought it might be useful for transmitting signals through the air. Gordon Gould thought it could concentrate light energy onto a small, intense spot. The Pentagon thought laser beams might guide weapons to their targets, or perhaps destroy enemy targets by themselves. But these were all vague possibilities. As Irnee D'Haenens pondered the ruby laser that he had helped create, he told Maiman that the laser was "a solution looking for a problem." It was a bright and promising newborn, but its ultimate potential was unknown.

THE BASIC IDEA OF THE LASER traces back to a theory that Albert Einstein proposed during World War I to explain how atoms absorb and release energy. Atoms have various amounts of energy, as if they were standing on different rungs of a ladder (Fig. 1.1). Those with the minimum possible energy are in the ground state, as if they were standing on the ground at the base of the ladder. Those with more energy are on higher rungs (left side of Fig. 1.1). The steps aren't evenly spaced on a real energy-level ladder, but the spacing is the same for all atoms of the same element.

The energy comes and goes in the form of light and other electromagnetic waves—radio waves, infrared light, microwaves, and so on. Atoms absorb or release one chunk or quantum of electromagnetic energy at a time (physicists call the chunk of energy a photon) and its energy depends on the frequency of the wave, the higher the frequency, the more energy. Absorbing a photon

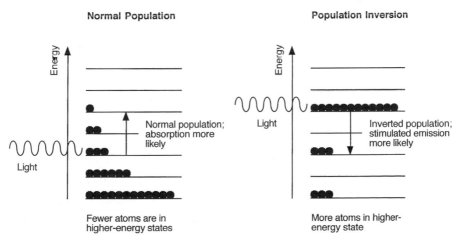

FIGURE 1.1. Normally the number of atoms in an energy level decreases with the energy. A population inversion occurs when more atoms are in a higher state than a lower one. In this diagram, as in real atoms, only certain transitions are possible.

increases the atom's energy, pushing it higher up the energy-level ladder. Releasing a photon decreases its energy, so it slips down the energy-level ladder. The process of jumping between rungs is called a transition, and it involves absorbing or releasing exactly the difference in energy between the two levels.

In 1916, Albert Einstein realized that an atom with extra energy could drop to a lower energy level in two different ways. One is spontaneous emission, where the atom makes the transition all by itself, releasing the excess energy as a photon. Like radioactive decay, the chance of spontaneous emission is measured by the lifetime of the excited state. If the high-energy state has a one-second lifetime, half of the atoms in that state will drop to the lower state within one second, and three-quarters will drop in two seconds. Most light we see in the everyday world, from the sun and artificial lamps, is spontaneous emission, and people knew about it long before Einstein.

Einstein added an alternative called stimulated emission, which can occur when the excited atom encounters a photon having exactly the amount of energy it would release by dropping a step down the energy-level ladder (Fig. 1.2). This causes a resonance that essentially tickles the excited atom, stimulating it to oscillate at the same frequency as the photon, and making it more likely to release a second photon. If the incoming photon stimulates emission, the atom will release a photon exactly in phase with the first. That is, the two photons have the same wavelength, and their peaks and valleys line up perfectly. (Well, *almost* perfectly because of the uncertainty principle, but let's not worry about that here.) Every transition has its own probabilities for stimulated and spontaneous emission.

To Einstein, stimulated emission was the flip side of absorption. When an atom absorbed a photon, it made a transition from a low-energy state to one with more energy. Stimulated emission caused an atom to drop from a high-energy state to one with lower energy. Absorption transferred energy to the atom; stimulated emission took it away. In theory, Einstein was talking about all types of electromagnetic waves, but he gave light as an example.

Although the idea of stimulated emission sounded interesting, it initially seemed only of theoretical interest. Richard Tolman, a prominent physics professor at Caltech, noted that the process, which, like Einstein, he called "negative absorption," could amplify a beam of light. However, he argued that the degree of amplification would be negligible. The reason was that atoms (or, equivalently, molecules) normally tend to occupy the lowest possible energy state. Only a few have more energy, and the higher the energy level, the fewer

Absorption
Atom in the ground state absorbs a photon and is excited

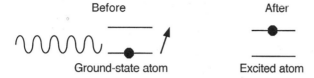

Before   After

Ground-state atom   Excited atom

Spontaneous Emission
Excited atom releases its energy on its own as a photon

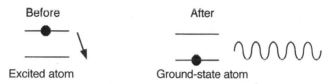

Before   After

Excited atom   Ground-state atom

Stimulated Emission
Excited atom is stimulated to emit energy, producing second identical photon

Before   After

Excited atom   Ground-state atom

FIGURE 1.2. Three possible transitions between a pair of energy levels. An atom in the lower state can absorb a photon and jump to the higher level, an atom in the higher state can spontaneously emit a photon, or an atom in the higher state can be stimulated to emit light.

atoms are in that state, as shown at left in Figure 1.1. The more abundant low-energy atoms would absorb photons before they could stimulate emission from the few high-energy atoms.

Yet only a few years later, Rudolf Ladenburg, working in Germany, indirectly observed stimulated emission. He was studying the optical properties of neon gas at wavelengths near a transition where it strongly absorbed and emitted light. Theory predicted that stimulated emission would change those properties, and when he fired an electric discharge through the neon, he saw the expected change. It was the first evidence that stimulated emission existed.

Ladenburg saw stimulated emission because he excited the gas, changing the populations of atoms in the excited states. Electrons hit the atoms, exciting them to a higher energy level, then the atoms dropped back to their normal lower energy state. The effect was like throwing marbles on a staircase; the marbles stayed briefly on the upper stairs, then dropped down to the bottom. Ladenburg had briefly put enough atoms onto a higher-energy step to see their stimulated emission. It wasn't enough to overwhelm the absorption by atoms in the lower state, but it partly cancelled the effects of absorption. Yet just as the marbles dropped to the floor, the excited gas atoms dropped back to their lower energy level and stimulated emission faded back into insignificance. Everything returned to what physicists call equilibrium—a state where the energy is balance, and the whole system doesn't change as long as it's left alone. Most everyday objects are at equilibrium, from the book in your hand to a hot cup of tea. A population inversion like at the right side of Figure 1.1 seemed only a brief permutation, not something of major importance.

Physicists liked the boring normality of equilibrium because it was easier to understand than a dynamic, changing system. They might venture away from equilibrium briefly for an interesting experiment, then return to comfortable normality. After Ladenburg saw the effect he wanted, he decided he was finished and stopped the experiment. If he had turned the power up higher, he might have excited many more atoms, producing a condition called a "population inversion," where more atoms were in a higher energy level than in a lower one. However, that would have taken the gas out of equilibrium into a much messier situation that physicists didn't understand very well. He might have seen something new, but he wouldn't have been able to tell what caused it, so the experiment would have been of little value.

Ladenburg was not alone. The years between the two world wars were a fertile time for studies of how atoms and molecules absorbed and emitted light, a field called spectroscopy. Specialists systematically measured the wavelengths of bright and dark lines in the spectra of gases, and from them deduced the energy levels and internal structures of atoms. German spectroscopists were particularly meticulous in compiling tables of spectral lines produced by exciting gases with electric currents, and once in a while they saw surprisingly strong emission. It's possible they excited the gases enough to produce stimulated emission a few times, but they didn't recognize it at the time because they were looking for equilibrium.

FIGURE 1.3. Valentin Fabrikant, the first person to envision optical amplification. (Courtesy of George R. Gamsakhourdia)

One of the few spectroscopists to spend much time wondering what might happen outside of equilibrium was Valentin Aleksandrovich Fabrikant (Fig. 1.3). Born October 9, 1907, Fabrikant came from a well educated Jewish family; his father was a professor of agriculture. Fabrikant studied at Moscow State University under two highly respected Russian spectroscopists, G. S. Landsberg and L. I. Mandelstamm. In the 1930s, he settled in to teach physics at the Moscow Power Engineering Institute, and did research on the optical properties of gas discharges at the prestigious Lebedev Physical Institute in Moscow. His research led him to wonder what might happen if the majority of gas atoms were excited, inverting the normal population of energy levels.

It was a question no one else had asked before. Trained to think the world was at equilibrium, they didn't wonder what might happen if more atoms were in the excited state than in the ground state of a transition. Fabrikant realized that the photons would then be more likely to encounter excited atoms and stimulate emission. If stimulated emission exceeded absorption, the number of photons would increase—amplifying the light passing through the gas. (Like Einstein and Ladenburg, he actually called the process "negative absorption.") He suggested a way to produce a population inversion in a doctoral dissertation he completed in 1939. But with World War II on the horizon, Fabrikant stalled.

Fabrikant didn't fit neatly into any of the roles outsiders expect to find in the Stalin-era Soviet Union. He was a modest, careful intellectual, whose work in physics kept him apart from political turmoil. Fabrikant was not socially active, neither part of the ruling Communist apparatus nor a persecuted dissident. His father was arrested in 1930, but was quickly released when no charges could be proved, not shipped to the gulag or executed like later dissidents. When the young Fabrikant visited his father in jail, he talked with a sympathetic young secret policeman who helped him get an academic post. Stalin's Great Terror came later, in 1937, and even at its peak it didn't throw all of Russia into the gulag.

Like many other Russians, Fabrikant didn't try to draw attention to himself. He settled into a secure post as head of the physics department at the Moscow Power Institute. It was not Moscow's top school, but it had some strong departments and some brilliant people worked and studied there. Among them was Andrei Sakharov, who worked under Fabrikant in the physics department before becoming involved in nuclear weapons. As the war raged, Fabrikant set aside his dissertation research to concentrate on teaching physics and developing practical fluorescent lamps, which produce spontaneous emission. The Soviet system didn't put a premium on publishing and promoting one's own work, and Fabrikant was building his reputation as an educator rather than a researcher.

Only after the war did Fabrikant have time to play with an idea as exotic as stimulated emission. He worked on it with a woman colleague, Fatima Butayeva, but later admitted he "did not pay attention to the practical value of this idea." The idea percolated in his mind, and he mentioned it a few years after the war at a lecture he gave at the Power Institute, where he substituted for an ill speaker. Michael Vudynskii, who attended the talk, was intrigued, and joined Fabrikant and Butayeva in filing for a Soviet patent in June 1951, which described how to amplify light and radio waves. Soviet patents were a way of garnering some recognition, but they carried no property rights, and the system proceeded leisurely. His application remained unpublished for several years, and Fabrikant reported no experiments until later.

American physicists also were caught up in the turmoil of World War II, which shifted their focus from research to the urgent task of winning the war. Among them was Charles Hard Townes, who would play a pivotal role in laser development.

Born July 28, 1915, in Greenville, South Carolina, Townes had roots tracing back to the Mayflower and early settlers of Virginia. He was raised on a 20-acre farm, owned by his father, a lawyer. The family valued education; Townes's siblings also went to college, his older brother became a noted entomologist, and an older cousin was a college professor. Charles took to science early, and focused his mind to speed through school, graduating from a local college at age 19. He spent a year at Duke University before heading west with $500 in savings to earn a doctorate at Caltech, where Tolman was one of his professors. Hoping for a professorship when he graduated in 1939, Townes initially was disappointed when he landed a job at Bell Labs, then in downtown Manhattan. However, he warmed to the place when he was allowed to do fundamental research.

That changed abruptly as America mobilized for World War II. Bell's research director summoned Townes to his office one Friday in early 1941, and told him he would start work on a military radar project the following Monday. Townes was not happy. He wrote in his autobiography, "it was a kind of dull and unattractive business. To be trying to think of ways to destroy things and kill people was not inspiring at all." Yet he had a keen sense of duty, and duly turned to radar development.

First developed in the 1930s, radar was a great advance in military technology because it spots otherwise invisible targets by detecting their reflections of coded signals from a radio transmitter. The first radars used radio waves about a meter long, at frequencies of hundreds of megahertz in today's broadcast television bands. However, those radio waves spread across a broad angle from the antenna, and couldn't discern much detail. Military planners wanted higher frequencies in order to get better sensitivity, tighter beams, and antennas small enough to mount on an airplane. Townes began working at three gigahertz, a factor of ten higher than early radars, but military planners soon asked for 10-gigahertz radars, then for systems operating at 24 gigahertz (GHz), corresponding to a wavelength of just 1.25 centimeters. The first two worked, but were never used in battle. The 24-GHz system ran into problems that helped point Townes toward the laser.

Air is remarkably transparent to radio waves because radio-frequency photons have too little energy to excite air molecules to rise a step up on their energy-level ladders. At higher frequencies, the photons carry more energy, eventually reaching a point where air molecules absorb them, thus climbing a small energy-level step. As for visible light and atoms, the interaction with microwaves revealed the internal structures of molecules. Physicists had begun exploring microwave spectroscopy in the 1930s. It was the sort of fundamental research that interested Townes, and he had noticed one early discovery. Ammonia molecules absorb microwaves at 24 gigahertz, exciting the molecule's single nitrogen atom to vibrate at that frequency. Ammonia is very rare in the atmosphere, but water molecules have a similar structure, and Townes found they also absorbed at 24 GHz. That caused the humid air over the ocean to absorb the 24-GHz radar signals, limiting the radar's range to a few miles, too short to interest the Pentagon.

When the war was over, Townes returned eagerly to basic research—in microwave spectroscopy. War work had taught him new skills and techniques, and the end of the war made a wealth of sophisticated new equipment available to researchers. Physics before the war was often a low-budget operation, with apparatus held together by string and sealing wax. When war labs shut down, their expensive equipment became war surplus junk. Military agencies were glad to give it to physicists, who were delighted to get sophisticated equipment they couldn't afford to buy. Microwave spectroscopy boomed.

The postwar physics boom brought a wave of discoveries. One was that the spin of an atomic nucleus tends to align along the same direction as a magnetic field; the stronger the field, the stronger the alignment. The atomic spins can point in either of two directions along the lines of the magnetic field, and the two states have slightly different energies. The atoms absorb or emit radio waves as they flip back and forth between the two states. Today it's the basis of magnetic resonance imaging. Then it gave postwar physicists a new way to think about energy levels, absorption, and emission. They found that magnetic fields affected the spins of electrons as well as of atomic nuclei. They measured the how long it took atoms in a higher energy state to spontaneously emit radio waves and drop to a lower energy state. As they learned more about atoms and their transitions between energy levels, they began to think the hitherto unthinkable: Perhaps by manipulating magnetic fields or other effects, they could create a state where more atoms or molecules were in a higher energy level than in a lower one.

The idea came to Willis Lamb Jr., a Columbia University physicist, after he showed that applying a magnetic field could split energy levels of electrons orbiting hydrogen atoms into pairs of closely spaced levels. Thinking further, Lamb realized that hydrogen would absorb radio waves on the transition between the two levels only if there were more atoms in the lower level than in the higher one. If more atoms were in the higher state, stimulated emission would overwhelm absorption, producing "negative absorption" of radio waves—amplification in ordinary language.

Fundamentally, what Lamb envisioned was a population inversion between two specific energy levels, separated by the energy of a radio-frequency photon. Fabrikant had envisioned a population inversion between a pair of energy levels separated by the energy of a visible photon. (The real importance of a population inversion is when the atom or molecule can make a transition directly between the pair of energy levels involved. Quantum mechanical rules prohibit transitions between some pairs of energy levels. The details aren't important, but you should know that only certain transitions are possible.)

Edward M. Purcell and R. V. Pound of Harvard used this effect to produce stimulated emission. First the researchers applied a magnetic field, and the atoms aligned their spin with the field so the atoms had the lowest possible energy state. Then they flipped the direction of the magnetic field, while the atoms kept spinning in the same direction, so the majority of atoms were suddenly in the higher-energy state. The energy difference was tiny, but they were able to observe stimulated emission at 50 kilohertz, a frequency corresponding to the energy difference between the two states.

They called the phenomenon "negative temperature," a term that seems bizarre to anyone who learned that nothing can be colder than absolute zero. The negative value came from an equation used in thermodynamics to calculate the relative number of atoms or molecules occupying a pair of energy levels if the temperature and energy difference are known. As long as the world is at equilibrium, with more atoms in the lower-energy state, so the equation gives a positive value for temperature. The only way to get the equation to show more atoms in the upper energy level is to use a negative number for the temperature. That doesn't mean that the material is colder than absolute zero; it's just an arcane consequence of the assumptions made in deriving the equation. Today physicists say they have a population inversion when more atoms are in the upper energy level than the lower one.

These discoveries made microwave spectroscopy a hot field. Bright, intense and ambitious, Charles Townes was in the middle of the action. As his reputation grew, Columbia University offered him the professorship he had always wanted.

It was a plum position. Columbia was a place of makers and shakers. In September 1947, it named war hero Dwight D. Eisenhower as its next president; he assumed the office the following May. The physics department was headed by I. I. Rabi, who had received the 1944 Nobel Prize and headed the Columbia Radiation Laboratory, which developed radar systems during the war. To help him recruit a world-class faculty, the Pentagon offered him $500,000 a year for department-wide research. That was a lot of money at the time, and Rabi said it was more than he needed, so the Army cut the offer. Nonetheless he had plenty of money to hire Townes as a full professor for $6000 a year at the start of 1948.

The military funding came with strings attached, but they were loose ones. Townes and his military sponsors shared a common interest in developing higher-frequency microwave sources. The Pentagon's interest was practical: building better communication systems and radars. Townes wanted to create higher frequencies in order to probe the inner workings of molecules in ways impossible at lower frequencies. He agreed to head a seven-man panel to advise the Navy on the high end of the microwave spectrum, called millimeter waves because the wavelengths are measured in millimeters and frequencies of tens of gigahertz.

The prospects didn't look very good when they first gathered in early 1950. Existing microwave sources fed energetic electrons into a cavity where they bounced back and forth emitting microwaves at least half the length of the cavity. That worked well at frequencies of a few gigahertz, where the resonant cavities are a reasonable size. Waves from a three-GHz source are ten centimeters long; waves from a 10-GHz source are three centimeters. But at 60 GHz, the waves are a mere five millimeters, and higher frequencies have even shorter waves. It's hard to pack the required power into the cavity without generating more heat than can be removed. The mechanical tolerances looked forbiddingly tight. Progress was elusive, and as the committee's 1951 meeting neared, Townes worried that it was getting nowhere.

2

MICROWAVES ARE THE FIRST STEP

CHARLES TOWNES was notorious among Columbia graduate students for arriving at the Pupin physics lab early in the morning, singing or whistling cheerily, before the cleaning woman had arrived or the students had staggered off to bed. His children were young, and he was the sort of self-disciplined man who rose early to make the most of the day.

Townes woke soon after dawn on the day of the Navy millimeter-wave committee meeting, April 26, 1951. The date and place were set to coincide with the American Physical Society's spring meeting in Washington. Townes slipped quietly out of the room at the Franklin Park Hotel he was sharing with Arthur Schawlow, a Columbia postdoctoral research fellow who had come to give a paper at the physics meeting. They had worked together for nearly two years and become friends, but Schawlow was a bachelor and used to sleeping late. (That was about to change; Townes had introduced Schawlow to his younger sister Aurelia a few months earlier, and they would be married in May of the same year.) The courtesy of leaving Schawlow to sleep was typical of Townes, a gentleman of the old school who blended the manners of his native south with a natural reserve and a dogged determination.

Townes sat on a bench in Franklin Park among red and white azaleas. He recalls it as "a beautiful morning," but beauty didn't divert his formidable power of concentration. Still searching for a millimeter-wave source, he thought of tiny vacuum tubes, then wondered about using molecules. The idea of molecular electronics had intrigued him for years. Perhaps he could put energy into the molecule, then concentrate the energy into a natural molecular resonance, and extract it in the form of millimeter waves at the resonance frequency. Microwave spectroscopy had shown him that atoms in a molecule vibrated at the right frequencies.

At first glance, this idea seemed in flagrant violation of the second law of thermodynamics. Loosely speaking, that law says that energy tends to disperse from places where it is concentrated. That implies that you can't concentrate energy in a high-energy molecular resonance because the energy would disperse to lower-energy states. Thus there always would be more molecules in low-energy states than in higher-energy ones—as long as the molecules were in thermodynamic equilibrium.

Townes didn't explicitly rule out equilibrium, but something on that bright morning set him on a different mental path. He realized it was possible to select the high-energy molecules. Rabi had already done that by passing molecules through a magnetic or electric field, which bent their paths in a direction that depended on their internal spin energy. Doing the sort of thought experiment made famous by Einstein, Townes imagined separating excited molecules from unexcited ones, so only excited molecules were left. In that case, the first molecule to release its energy by dropping to the lower level could stimulate emission of an identical photon from a second excited molecule. Then those two photons could stimulate emission from other excited molecules. With no molecules in the lower energy level until the process got started, there would be nothing to absorb the growing cascade of photons. Pick the right resonance and you could generate the frequency you wanted. By escaping equilibrium, Townes had found a loophole in the second law.

Stimulated emission hadn't interested Townes before because it wasn't at the cutting edge of research. The idea of using it as a microwave source changed that. Rabi's molecular beam techniques could select the excited molecules he needed. Inspired by the idea, Townes pulled an envelope from his pocket and calculated how well the idea would work for ammonia molecules switching between two rotational energy states. He had worked with ammonia before, and knew the molecule so well that he had memorized its key properties. The wavelength was about 0.5 millimeter, corresponding to a frequency of 600 GHz, in the range that interested the Navy.

The source he envisioned was an oscillator, which generates electromagnetic waves at a specific frequency, like a single pure musical note. He recalls that his rough calculations showed "it was just marginally possible that the idea would work with ammonia." The beam of ammonia molecules itself could not produce enough energy by stimulated emission. However, the microwave energy could build up to the desired power if it passed through a suitable chamber with reflective walls. Engineers call this a resonant cavity, and it's essentially an echo chamber for electromagnetic waves that fit exactly into the cavity. Townes realized that stimulated emission from the ammonia molecules would build up as the waves bounced back and forth in the cavity, oscillating at the frequency he wanted.

Townes described the idea to Schawlow back in their room, but said nothing to the Navy committee. It wasn't the right forum for what he knew was a half-baked idea. Back at Columbia, Townes hauled out his notes from a Caltech quantum-mechanics class and took a closer look at stimulated emission. He found that stimulated emission could amplify a signal at the right frequency as well as generate a signal in an oscillator. Fabrikant had recognized amplification was possible a decade earlier, but he had stalled before thinking of oscillation or resonant cavities. Townes may not have enjoyed developing microwave sources during the war, but it had introduced him to the resonant cavity, which reflected signals back into the zone where they could be amplified to build up power. That insight would prove essential.

Townes's old quantum mechanics notes also pointed out something else—a stimulated emission photon has the same wavelength, or equivalently, frequency —as the photon that stimulated it. That was good news for a spectroscopist. It meant that the microwaves produced by stimulated emission would be concentrated in a very narrow band of frequencies. The narrower the line width, the more details a spectroscopist could see.

On May 11, Townes wrote up the idea in his laboratory notebook, an important formality to demonstrate priority if he wanted to patent it. He sketched a beam of ammonia molecules passing through a slit, with a magnetic or electric field deflecting the low-energy molecules, and the high-energy ones passing into a resonant cavity with a small hole to let some microwave energy leak out, shown in fig. 2.1. He carefully dated it and signed it "Chas. H. Townes." The following February, Schawlow signed it with the notation "signed and understood," adding that Townes had explained the idea to him on April 26, 1951.

Townes was well aware that others were thinking about stimulated emission. Looking back, he reflected, "The whole plan only required properly putting

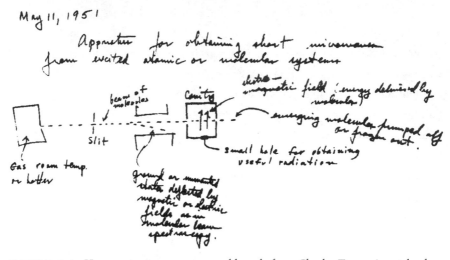

FIGURE 2.1. How a microwave maser would work, from Charles Townes's notebook. A magnetic or electric field would remove the unexcited ammonia molecules, directing only the excited ones into the cavity. (From *How the Laser Happened: Adventures of a Scientist*, by Charles H. Townes, copyright 1999 by Oxford University Press, Inc. Used by permission of Oxford University Press, Inc.)

together a number of ideas that were already known and floating around. Also critical was the recognition that this plan could be important."

Settling down and doing something may have been the crucial factor for Townes's success. He had other projects in the works, including another type of microwave source that at the time seemed more viable. Yet he was a careful and deliberate man, not about to let go of a promising idea. He saw the new microwave oscillator as a good project for a gifted graduate student, but first he had to convince himself the idea had a chance of success. Others never got that far. Purcell and Pound at Harvard saw stimulated emission at 50 kilohertz in 1951, when they flipped the spin of atomic nuclei, but they called it negative temperature and wandered away to other things. Willis Lamb and Robert C. Retherford at Columbia had noted in 1950 that stimulated emission could amplify, but like Fabrikant they did nothing further. Townes himself had thought of molecular-beam amplification as far back as 1948. About a year later, a young postdoctoral research fellow at Columbia, John W. Trischka, came up with the same idea independently and briefly pursued it before giving it up as too difficult. None of them had thought of a resonator.

Townes talked with others at Columbia, refining his ideas for a dissertation project. He recommended it to Jim Gordon, a third-year graduate student from MIT who had experience with molecular beams, and who had finished most of his Columbia course work. Townes warned Gordon that he might not produce a microwave source, but said he should get enough spectroscopic data to write a thesis. That would avoid the graduate student's nightmare: wasting a couple of years on a professor's hare-brained scheme.

Townes also had money to hire a recent Ph.D as a research assistant who could help Gordon. Schawlow had held the post, but his time ran out after Townes introduced him to his sister Aurelia, who was studying music in New York. The two fell in love and married just three months after they met. In gaining a brother-in-law, Townes lost a colleague, because university nepotism rules banned them from working in the same department. He filled the slot with Herbert Zeiger, who had just finished a thesis involving molecular beam research under Rabi.

Preliminary calculations led to one important change. They shifted from the 0.5 millimeter rotational transition of ammonia to a vibrational transition at 1.25 centimeters. The much lower frequency of 24 GHz looked more promising, and was easier to work with than the original 600 GHz frequency, for which instruments were not readily available. The prospects for oscillation still looked marginal, so they needed all the help they could get. "It didn't seem obvious to anybody that this was going to revolutionize the world at the time it happened. There wasn't any great excitement over it," Gordon recalled later. Townes included the "molecular oscillator" project in the Columbia Radiation Laboratory's quarterly progress report in December 1951, and framed his notebook entries so he could apply for a patent if the idea became promising, but he didn't try to publish a theoretical paper to stake out the intellectual turf.

Others were beginning to see the possibilities. After hearing a lecturer describe stimulated and spontaneous emission, Joseph Weber said, "It was immediately obvious to me that the stimulated emission idea could be used to make a microwave amplifier and generators." While teaching electrical engineering at the University of Maryland, the former Naval officer sat down and developed a plan for a microwave amplifier in 1951. He described the idea at a June 1952, vacuum-tube conference in Ottawa, where he impressed Rudolf Kompfner, a top microwave engineer at Bell Labs. Yet like Fabrikant, Weber had thought only about making an amplifier, not an oscillator. Weber did not mention a resonant cavity and lacked money to build anything.

Columbia had money for hardware, which Zeiger and Gordon spent a year designing and building in a laboratory on the 11th floor of the physics building. Townes kept in touch, but was not heavily involved. His time was stretched thin, although he worked long hours six days a week, saving Sunday for church and family. He supervised a dozen other graduate students, was involved in other research projects, was writing a book with Schawlow, and had teaching and administrative responsibilities.

Progress was slow. Calculations were time-consuming because they had to be done by hand or on electro-mechanical calculators, enhanced adding machines that ground out numbers with arrays of wheels and gears. Molecular beam systems were notoriously difficult to operate, requiring constant and careful attention. Zeiger left when his two-year fellowship ran out in early 1953, but Gordon plugged on.

Dwight Eisenhower cut military research spending after arriving in the White House in 1953, and those reductions strained the programs that supported the Columbia labs. Something had to go, and Rabi decided it was the molecular oscillator project, which already taken two years and consumed about $30,000, a lot of money for the university at the time. He and Polykarp Kusch marched into Townes's office to discuss the matter. They were a formidable pair, eminent professors confident of their opinions. Combative by nature, Rabi was a Nobel laureate. Some felt he had a nasty streak, and the graduate students feared him because he ruthlessly winnowed their ranks to select the best of the best. Kusch was less intimidating, but famed for his booming voice; he would win his own Nobel in 1955. Both had worked with molecular beams more extensively than Townes. They told him, "It isn't going to work. You know it's not going to work. We know it's not going to work. You're wasting money. Just stop!" Townes held steadfast. "Fortunately I had tenure," he recalled.

Townes was also fortunate that Jim Gordon was beginning to make headway. At the end of 1953 Gordon finally saw evidence of stimulated emission and amplification. Yet oscillation remained elusive.

The problem was a simple but stubborn one—balancing the flux of both ammonia molecules and microwaves into and out of the cavity, which was essentially a copper tube with holes in each end. They needed to keep microwaves building up in the cavity while letting the ammonia molecules pass through it. Only a small fraction of the microwaves should emerge. Opening the holes to admit more ammonia molecules allowed too many microwaves to leak out. Gordon and Townes tried putting rings into the holes for the ammonia molecules to block microwaves from escaping, but made little progress.

A breakthrough came in early April 1954, when Gordon pulled out the rings to open the ends almost completely, allowing more ammonia molecules to enter.

That increased the microwave intensity by a factor of ten. "It was that factor of ten we needed to make the thing oscillate," recalled Gordon. He burst into a seminar Townes was holding to report the good news. The group adjourned to the laboratory, first to check the evidence and then to celebrate. Townes announced their success publicly on May 1, in a late paper at an American Physical Society meeting. He and Gordon also dashed off a short paper to report their success in Physical Review. The output was a mere 10 billionths of a watt, but it was enough to show it worked.

Afterwards Townes sat down with his students over lunch to try to name the thing. They started with Greek and Latin words, but nothing worked, so they tried creating a descriptive acronym. They finally settled on Microwave Amplification by the Stimulated Emission of Radiation, which spelled out the name that would stick, MASER.

The first maser oscillator demonstrated the concept and gave Gordon enough for his thesis. Yet a second was needed to measure exactly how wide a range of frequencies the maser emitted, a concern that had grown increasingly important. Narrow emission was good for spectroscopy, and essential for another potential application—atomic clocks, which measure time by counting how many times an atom or molecule oscillates at a characteristic frequency. The more accurately the frequency is known, the more accurate the clock. Townes had long consulted for Harold Lyons, who was developing atomic clocks at the National Bureau of Standards. Lyons had already developed one that worked with cesium, but Townes hoped the maser might keep better time—if the ammonia molecules were emitting exactly the same frequency. Checking how accurate the frequency was required a second maser to compare with the first, but Gordon had other priorities. He wanted to finish his thesis and to take a six-week vacation he had planned in Europe. Not until November did they operate the second maser—shown in figure 2.2—and prove the emission was on as narrow a range of frequency as they had hoped.

In January, Columbia issued a press release touting the newly invented maser for its potential to make atomic clocks that kept time more accurately than the rotation of the Earth. The release did not reflect a real shift in goals. The public knew nothing about microwave spectroscopy, but Lyons had skillfully promoted his cesium atomic clock as a neat new technology, and the popular press had embraced the idea of atomic clocks. Columbia sought to capitalize on that preexisting interest.

FIGURE 2.2. Charles Townes and Jim Gordon show off the second ammonia maser they made at Columbia. T. W. Wang is in the background with the first ammonia maser. (From *How the Laser Happened: Adventures of a Scientist*, by Charles H. Townes, copyright 1999 by Oxford University Press, Inc. Used by permission of Oxford University Press, Inc.)

Meanwhile, Gordon and Townes had to finish their work on the maser. Gordon wanted to finish his thesis; Townes wanted to patent the idea and publish a long analytical paper. Both goals required exhaustively analyzing both maser theory and the experiments, and covering the details they had glossed over in their first short report. The task took months.

Before that work was finished, Townes received an unpleasant surprise. While Gordon ground away on their paper, Townes took a break in early April 1955, to attend a meeting of the Faraday Society in Cambridge, England. He had wanted to talk about the maser, but the conference organizers asked him to talk instead about magnetic effects in molecules. Townes duly did so, and was flabbergasted when Soviet physicist Alexander Prokhorov came to the podium and started describing the theory of an ammonia maser.

The subject of Prokhorov's talk hadn't been on the program. In fact, Prokhorov's very appearance had been uncertain. During the Cold War, conference organizers could never be sure if Soviet scientists would show up, and the Russians usually didn't announce the topics of their papers in advance. Townes had never met Prokhorov, and was only vaguely familiar with his work on microwave spectroscopy. But Townes would not sit still when someone else was claiming his invention. After Prokhorov finished, Townes stood and announced, "Well, that is very interesting, and we have one of these working." Then he described his experiments.

Prokhorov could not have been surprised, because he had read Townes and Gordon's brief description of their maser experiment well before the meeting. However, that short paper lacked a theoretical analysis of how the maser worked, which Prokhorov had provided. Townes and Gordon didn't submit their analysis until after the conference (*Physical Review* received it on May 4), and when it appeared in August it cited a paper by Prokhorov and his former student Nikolai Basov.

When the session was over, Townes and Prokhorov walked around Cambridge, comparing notes and getting to know each other. Prokhorov was an unabashed Communist; few who weren't were allowed to leave the country during the Cold War. However, he was an unusually friendly sort for a Soviet-era Russian. He also was fluent in English; he had been born July 11, 1916, in Australia, where his Russian revolutionary parents were in exile, and his family had lived there until 1923. He also had a solid knowledge of microwave spectroscopy that made the conversation fruitful. A month and a half after going back to the Lebedev Physics Institute in Moscow, Prokhorov and Basov got their own ammonia maser working.

Townes and Prokhorov found themselves in the first of what would become many hard-to-resolve clashes over priority. Both groups had been working for years on their ideas. Townes had his notebook record from 1951, and the Columbia progress report from the end of that year. The Russians said they first discussed their ideas at the May 1952 *All Union Conference on Radio Spectroscopy*. However, their first formal publication was not until the October 1954 issue of the Russian-language *Journal of Experimental and Theoretical Physics,* the Soviet counterpart of *Physical Review*. That paper considered not ammonia but cesium fluoride, a molecule they decided wouldn't work, but it did describe maser theory. Townes and Gordon clearly made the first working maser, but Prokhorov and Basov could claim they first published a theoretical analysis of how the maser worked. Both achievements count in physics.

It's impossible to definitively sort out the details. Americans at the time claimed the Soviet establishment exaggerated their scientific achievements, and doubtless there was some exaggeration. Yet Russians also made substantial contributions to physics, particularly in theory. Russian physicists tended to publish less, and wrote almost exclusively in Russian, a language few Americans could read. Some Americans suspected that the Russians may have had access to the Columbia quarterly reports, and that the 1952 "All Union" conference may never have happened. It's probable the Russians picked up a few ideas; science works that way. But it also seems clear that much of their work was independent.

Basov and Prokhorov soon took another significant step beyond the original ammonia maser. Townes produced a population inversion by physically separating the excited ammonia molecules from low-energy molecules. That worked for a few systems, but was cumbersome. Basov and Prokhorov proposed a maser involving three energy levels instead of two. Their idea was to excite most of the gas atoms or molecules from the ground state to the highest of the three energy levels. This could produce a population inversion involving the middle state in two ways. Atoms could be excited to the upper state and accumulate there while the middle state remained empty, creating an inversion between the top and the middle state. Or most of the atoms could be excited from the ground state to the top state, then drop to the middle state, creating an inversion between the populated middle state and the depleted ground state. Fig. 2.3 shows the two possibilities. The details were vague, and the idea never quite worked exactly as Basov and Prokhorov proposed, but the idea opened a new door.

The maser soon caught the attention of more physicists because it promised interesting possibilities. Spectroscopy was an obvious possibility. Atomic clocks

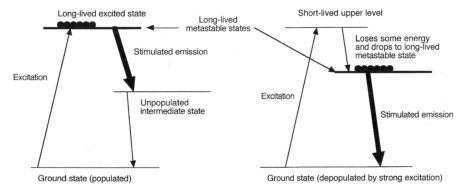

FIGURE 2.3. Three-level masers can operate in either of two ways, with the laser emission on a transition between the upper and middle levels (left) or between the middle and lower levels (right).

FIGURE 2.4. Prokhorov, Townes, and Basov (left to right) when Townes was visiting their microwave maser laboratory at the Lebedev Physics Institute. (Courtesy of Dimitri Basov)

interested military planners as well as physicists, because precise clocks could help guide intercontinental ballistic missiles to their distant targets. The maser also promised the ability to amplify faint microwave signals with extremely low noise, important for both research in radio astronomy and military radar systems.

As new people started entering the field, Townes wandered away. After seven years on the Columbia faculty, he took his first sabbatical in the summer of 1955. The ideal of the sabbatical is to give professors time to recharge their intellectual batteries, and Townes saw it as a chance to explore. He set out for laboratories in Europe and Japan, and visited Prokhorov and Basov in their Moscow lab as shown in Fig. 2.4. But he was looking to stay on the frontier of physics, not linger in a field that was maturing. He had done the maser; he had a year to look for something new.

Among the new entrants was Nicolaas Bloembergen, a Harvard physicist who had grown up in Holland and been educated in elite Dutch schools. He survived World War II by dodging German soldiers and eating bitter tulip bulbs to fill his stomach, and was eager to finish his schooling outside of ruined Europe. Brilliant and intense, he pioneered research on the effect of magnetic fields on the spins of atomic nuclei with Pound and Purcell, and relished the intellectual challenge of the complex interactions of atoms in solids.

Initially Bloembergen had little interest in population inversions or masers. Everyone else was talking about gases, where the molecules are far apart and interact in simple ways with each other, producing the sharp energy-level transitions needed for atomic clocks. He worked with solids, where the complex interactions of adjacent atoms spread energy-level transitions over too broad a range of frequencies for atomic clocks. Bloembergen didn't see any use for solid-state masers until May 17, 1956, when Woody Strandberg of MIT told him population inversions in solids might be useful as low-noise microwave amplifiers and receivers. That set Bloembergen systematically considering ways to take advantage of the rich range of interactions possible in a solid.

Inspiration struck on June 12, and just under two weeks later Bloembergen handed the typist a draft of his design for a three-level maser. It was entirely independent of Basov and Prokhorov's proposal, which he hadn't seen, and went well beyond it. Bloembergen started by considering an atom that could occupy many possible energy levels in a solid. Earlier experiments had shown that pumping a lot of energy into the system at one transition—a process called "power saturation"—could cause population inversions between pairs of other energy levels. Bloembergen analyzed the possibilities for three energy levels, but made it clear that his ideas applied if the system had more energy levels. He realized it was important to consider how quickly atoms spontaneously dropped from high-energy states on their own, without stimulation.

After describing the general idea, Bloembergen focused on specific materials that he thought might work when a magnetic field interacted with electrons spinning in the solid. Electrons occur in pairs that spin in opposite directions, but which have the same energy when no magnetic field is present. Adding a magnetic field increases the energy of electrons spinning in one direction, and decreases the energy of electrons spinning the other way. Thus it splits what looks like a single energy level into two distinct ones. If he cooled a solid close to absolute zero, put it into a magnetic field, and illuminated it with microwaves tuned to excite atoms to the right energy level, he figured he could make a solid-state maser. It was brilliant step forward that drew on his earlier nuclear-resonance research.

Bloembergen tried to build his own maser operating at the 21-centimeter line of interstellar hydrogen. Later in the summer, Derrick Scovil came up with essen-

FIGURE 2.5. Nicolaas Bloembergen with the three-level solid-state microwave maser he demonstrated at Harvard in March 1958 (Courtesy of Harvard University and Nicolaas Bloembergen)

tially the same idea independently at Bell Labs. Bell had greater resources, and Scovil picked an easier material, so in December 1956, Scovil beat Bloembergen to make the first three-level maser and the first solid-state maser, which operated at 9 GHz. Bloembergen later made his own maser, shown in figure 2.5.

Both the three-level design and the use of a solid were vital steps if masers were ever to become practical. The two-level ammonia maser was important because it demonstrated the maser concept, but it could go no further than the laboratory. Like other atomic and molecular beam machines, it was large, complex, and sensitive. That was fine for graduate-student experiments, but limited its potential uses to a few highly demanding laboratory applications such as extremely precise atomic clocks. The solid-state maser was far more practical.

Solid-state masers made from a variety of other materials followed in short order. A clear winner emerged at the end of 1957, when Chihiro Kikuchi demonstrated a maser made of synthetic ruby at the University of Michigan's Willow Run Laboratory. Made by growing crystals of aluminum oxide containing a small fraction of chromium, synthetic ruby was readily available, durable, and easy to use.

Military scientists and engineers saw the maser as a potentially valuable low-noise amplifier for microwave signals, and military contracts fuelled a maser boom. Willow Run was a military lab run by the university. Well-established aerospace and electronic giants such as Westinghouse and the Radio Corporation of America jumped into masers. So did upstarts like the fast-growing Hughes Aircraft, which launched its effort in 1955 by hiring atomic clock developer Harold Lyons to start a maser research group.

As the maser matured, its limits became apparent. Solid-state masers required cooling to temperatures near absolute zero to produce a population inversion on the low-energy microwave transition. Gas lasers remained too sensitive for use outside a precision laboratory. A cynic redefined MASER as "Means for the Acquisition of Support for Expensive Research." In the end, other microwave technologies would eclipse the maser for all but a handful of uses.

Yet the maser did show that stimulated emission could be tamed to generate electromagnetic radiation. It set some people dreaming of generating a different kind of electromagnetic radiation with much shorter wavelengths and higher frequencies—what we call light.

3
LEAPING A FEW ORDERS OF MAGNITUDE: THE OPTICAL MASER

THE MASER GOT PHYSICISTS THINKING about stimulated emission. No longer was it a physical curiosity of purely academic interest. It was a effect that could generate and amplify microwaves. The maser's microwave emission was confined to a very narrow range of frequencies, making it very attractive for spectroscopy and atomic clocks. As a device that amplified stimulated emission, it was a step on the road to the laser.

Yet the microwave maser remained fundamentally a microwave device. The cavity in which the microwaves resonated was roughly the same size as the microwaves themselves. It usually directed its output into a hollow metal tube called a "waveguide," roughly the width of a single wavelength, which delivered the microwave energy to measurement instruments or test cells. Although the laser was based on the same physical principles as the maser, it was a long way down the road.

Stimulated emission is a general process which occurs across the entire electromagnetic spectrum. As early as 1950, Pound and Purcell saw stimulated emission at 50 kilohertz, a low radio frequency. Townes considered an ammonia transition at 600 gigahertz, before deciding it would be easier to work at 24 GHz. That factor of 10-million range may seem large, but the electromagnetic spectrum has no limits. Extremely low frequency waves are thousands of kilometers

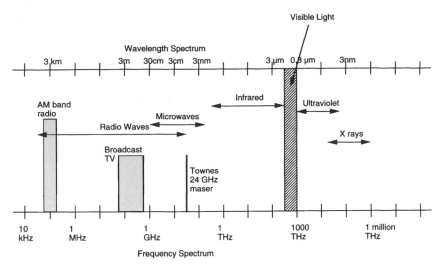

FIGURE 3.1. The electromagnetic spectrum.

long, oscillating a few dozen times a second. X rays oscillate at rates to 10^{18} (a million trillion) times a second and are as short as 0.3 billionth of a meter (0.3 nanometer). Gamma rays go to even higher frequencies. The more familiar realms of the infrared, visible light, and the ultraviolet are between microwaves and X rays (see fig. 3.1).

The maser extended stimulated emission from low radio frequencies to microwaves. For decades, electronic technology had steadily advanced from lower to higher frequencies, from kilohertz to megahertz, and from megahertz to gigahertz. It was only natural for physicists to think about pushing stimulated emission to even higher electromagnetic frequencies. The logical next step would be to frequencies above 100-GHz range, called millimeter waves because their wavelengths are measured in millimeters. Those frequencies were hard to generate in other ways, so the maser might help open a new part of the spectrum.

Yet that was easier said than done because vital technical infrastructure didn't exist. Electronic devices generally didn't respond above 100 GHz, so measurements were hard to make. Not much was known about frequencies from 100 GHz to about 100,000 GHz. Much of what was known was not encouraging. Molecules in the atmosphere strongly absorbed in much of that band, as they had blocked the 24-GHz radar Townes had developed during World War II.

The better-known realms of infrared and visible light beckoned beyond the *terra incognita*. Light frequencies are more than a thousand times higher than the microwave band, but physicists had worked with light for generations. Spectroscopists in the first half of the twentieth century had carefully measured the wavelengths where atoms and molecules emitted and absorbed light. Theoreticians had elegantly explained how the single electron of the hydrogen atom absorbed or emitted light as it made transitions between energy levels. Light was a territory so well explored that some physicists found optics boring.

As imaginative physicists learned how the maser worked, it was only natural for them to wonder about masers at infrared and optical wavelengths. Robert Dicke, a Princeton University physicist, jotted down some ideas about infrared versions of the maser in February 1956, which in retrospect were farsighted. He later received a patent. But most just thought and talked about it, because it was far from obvious how to cross the vast gap in frequency.

Townes came back from his sabbatical with his interest in masers revived. His first priority was to develop solid-state masers for radio astronomy, a young field that fascinated him. He thought about moving to higher frequencies, but couldn't come up with any good ideas. When Bill Bennett, a Columbia grad student, asked him about press reports of a light amplifier, Townes said they were probably mistaken. Townes turned out to be right about that, but wrong when he added, "It would be much too hard to build a real light amplifier." Townes also turned away William Ottig, a physicist at the Air Force Office of Scientific Research, who tried to interest him in infrared and optical masers in early 1957. But by the summer he had become cautiously optimistic, and told an Air Force study group meeting that an infrared laser operating at a wavelength of 10 micrometers (0.01 millimeter) might be possible in 25 years.

In late summer, Townes finished his part of the radio astronomy project, shown in figure 3.2, and turned to serious consideration of prospects for higher-frequency masers. He was a careful and methodical physicist, who thought first and systematically analyzed a problem before starting a project. He wanted to be sure an experiment would work before he tried it. It's the style of a physicist who believes his part of the world is orderly enough for him to calculate how it should behave. That approach had worked for him before, so he used it to approach higher frequencies.

His initial thought was to try incremental increases in frequency, but he found problems standing in his way. The size of a microwave cavity depends on its wavelength. The shorter the wavelength, the smaller the cavity, and the fewer atoms or molecules it could hold, so it would generate less energy and might fail

FIGURE 3.2. Charles Townes with a ruby microwave maser amplifier developed for radio astronomy. (Courtesy of Bell Labs)

to oscillate. Higher frequencies also require higher transition energies, posing another problem. The higher the transition energy, the faster the energy tends to leak away as spontaneous emission, which requires pouring more energy into the system to maintain the population inversion that allows stimulated emission. The effect is very strong—double the frequency or transition energy, and the rate of spontaneous emission increases by a factor of 16. The lack of good equipment to measure higher frequencies accurately posed more problems.

Townes decided it would be simpler to make the big jump over the vast *terra incognita* of largely unexplored frequencies to the better-known turf of infrared and visible light. The jump was a bold one, roughly a factor of 10,000 in frequency. Yet Townes felt encouraged when he analyzed the prospects for what he decided should be called an optical maser, because it was based on the same principle as the microwave maser. The energy-level transitions that produce visible and infrared light differ from those that produce microwaves. Visible light comes from large shifts, when electrons jump between different orbital shells. Microwaves come from much smaller changes, when an electron changes the direction of its spin, or when a molecule changes how it rotates. The difference makes it easier to get light in and out, and sometimes traps energy in certain states for unusually long times. That meant energy didn't leak away as quickly as he had first thought it might.

Townes was not an optics expert, but he knew a wealth of information was available on optical spectroscopy. For many years, other physicists had tabulated the wavelengths of visible and infrared light that various materials absorbed and emitted. Researchers had analyzed that data to tabulate the electronic energy levels of atoms. New spectroscopic tools were emerging. On his sabbatical, Townes had worked with Alfred Kastler at the École Normale Supérieure in Paris, who had developed a new way to excite materials for microwave spectroscopy. Kastler's trick was called optical pumping, and involved illuminating a gas with light chosen to excite it to certain energy levels that could produce the desired microwave transmission. It was an important research tool that would earn Kastler the 1966 Nobel Prize in Physics. Townes had earlier thought about optical pumping of microwave masers. Now he realized that optical pumping might also excite the optical energy levels he needed for the optical maser.

As his ideas started to crystallize, Townes wrote them in his notebook on September 19, 1957. He also asked questions of other physicists at Columbia, which Rabi had succeeded in making an intellectual hotbed. During the 1950s, ten men who would later receive Nobel Prizes served on the Columbia faculty. Ten other future Nobel laureates were Columbia students or research associates during that period. Townes networked well, gradually gathering more ideas and deepening his analysis.

On October 4, 1957, the Soviet Union shocked the American scientific establishment by sending the first artificial satellite into orbit. The "beep, beep, beep" from the orbiting Sputnik 1 threw a bucket of cold water in the faces of overconfident Americans. The shock sent politicians to decry the "missile gap" and the Pentagon to reorganizing its research programs. Budgets, cut after Eisenhower took office, soon got a boost.

Townes continued methodically with his research. In late October, he invited into his office Gordon Gould, a graduate student working under Kusch. Rabi had introduced Gould to optical pumping, and Gould had put it to work to excite thallium atoms in an atomic beam—the first optical pumping experiments in the U.S. Thallium is a heavy metal between mercury and lead on the periodic table, which has strong emission lines. Gould had built a lamp that vaporized thallium and excited the atoms to emit bright ultraviolet light, which could excite other thallium atoms in an atomic beam so he could measure their energy levels. He had thought about using his thallium lamp to optically pump a microwave-emitting maser, and months earlier had asked Townes about how to patent the idea.

Investigating light sources that might work for optical pumping of an optical maser, Townes asked Gould about the thallium lamp. Gould gave him the details he wanted, and the discussion turned to developing new types of masers. Townes said he thought masers could be made to emit visible light, and Gould agreed. He'd been thinking about the idea, too. They talked, and a few days later Gould gave Townes more data. Townes wrote in his notebook that he had talked with Gould on October 25, the day Sputnik 1 went silent, and October 28.

In fact, others at Columbia must have been thinking and talking about optical masers as well. The idea was in the air in the physics department. Townes had been talking about the possibility, and others who had been working with microwave masers also may have wondered about. It was the sort of far-out idea that physicists will talk about among themselves, then drop to go back to their other projects. Physicists are little different from other people in that way; life doesn't leave enough time to explore all the ideas that come their way. While others talked, Townes settled down to work on the optical maser. He didn't realize that Gould also took the idea very seriously, and was about to do the same.

It was a natural mistake for Townes. Gould had been around Columbia since 1949, and at 37 he was closer in age to Townes than to most of the other graduate students. He was bright enough, but the faculty saw him as an erratic worker, with a tendency not to finish projects. By the fall of 1957, Gould had largely finished his thesis research and was supposed to be working on his dissertation. However, he was bored with that process, and the conversation with Townes sent his mind racing in a more exciting direction.

Townes didn't notice because he was also turning in another direction, toward Bell Telephone Laboratories. After returning from his sabbatical, he had agreed to consult with Bell a couple of days a month. Bell was interested in masers for communications; its staff was laden with Columbia graduates including Jim Gordon, and Townes had his own personal ties. His friend, colleague, and brother-in-law Art Schawlow worked there. Schawlow's boss Al Clogston asked Townes if he could find something to inspire him. Townes wanted to help, so over lunch he told Schawlow he was thinking about optical masers.

Schawlow had wanted a university professorship when he left Columbia, but settled for Bell because his wife wanted to continue her singing lessons in New York City. Schawlow worried about ending up in a mundane engineering job, but instead was offered a slot as an experimental physicist working on superconductivity with John Bardeen, a leading theorist soon to share the Nobel Prize as co-inventor of the transistor. Schawlow knew little about superconductivity, but it was fundamental physics and Bardeen was well respected. However, Bardeen left for the University of Illinois before Schawlow arrived, leaving no one to guide him in the new field.

Warm and jovial but a bit shy and unsure of himself, Schawlow was not as intense or tightly focused as Townes—whose family said he once took a physics textbook with him to the circus. Schawlow was born in the United States, but raised in Toronto, where his mother's family came from. The family said little about their origins, and he didn't learn that his Latvian immigrant father had been born a Jew until he was 17. A bright boy fascinated by electronics, the young Schawlow initially wanted to be an engineer, but he was too young to qualify for an engineering scholarship when he graduated from high school at 16 in 1937, so he wound up a physics major. Schawlow liked to visualize things, which led him to the experimental side of physics, and in 1949 he earned a doctorate in optical spectroscopy at Toronto before spending two years at Columbia. Schawlow also developed a serious interest in jazz, and over the years accumulated a large record collection.

At Bell Schawlow finished writing a book on microwave spectroscopy with Townes, but had a hard time finding his direction in solid-state physics. He passed up a chance to work on masers once they got off the ground, because he felt out of touch. When management appointed him the department's safety representative, Schawlow suspected it was a hint they thought he had nothing better to do. "About the only thing I did was that I had to write an accident report when one of the theoretical physicists stabbed himself with a pencil—a sharp pencil," he recalled much later. His playful side couldn't resist having fun with the incident. He suggested that "theorists should be instructed on the use of pencils." It was a clever play on the difference between his role as an experimental physicist who understood how things worked and the theorist's comparative estrangement from reality.

The optical maser idea caught Schawlow's attention. He also had thought about making infrared masers, although only in the part of the infrared at wavelengths just shorter than microwaves. He and Townes talked a while, and decided to collaborate. It was a natural partnership. Townes, tall, serious, self-driven, and six years older, was the natural leader; Schawlow, shorter, rounder, and often rumpled, had worked well with him on the microwave spectroscopy book, which became a classic in the field.

Their partnership made the optical maser a Bell Labs project, so Townes stopped discussing it at Columbia, and shifted his work there to other projects. He always played by the rules, and to him that meant any patents they generated should belong to Bell. Townes gave Schawlow a copy of his notes, and the two began working regularly on the idea, which he insisted on calling an optical maser.

The concept of a short-wavelength maser remained in the air at Columbia. It was too big and too obvious an challenge for a collection of top-flight physicists to ignore, especially since Columbia was a hotbed of microwave maser research. The allure of the idea was so bright that it caught the eyes of many other physicists around the world. It spread at places where microwave masers were studied, like the Hughes Aircraft Company's research laboratories in California and the Lebedev Physics Institute in Moscow. Physicists working on microwave masers scratched their heads and wondered about optical wavelengths, sometimes jotting down notes about their ideas.

Townes and Schawlow thought they had a head start in leaping across the vast gap between microwaves and light. Talking about new challenges is easy; the hard part is doing something about them. Increasing the frequency by a factor of 10,000 was obviously ambitious, but as Townes and Schawlow played with equations describing the energy levels, they found it wasn't as bad as it had looked initially. They confirmed that spontaneous emission drains energy faster from more-energetic transitions, but found it is offset by a factor called oscillator strength, so the shorter-lived transitions produce stimulated emission—gain—more efficiently. At optical wavelengths, the two factors roughly cancel out for many possible transitions.

Their next step was to consider what materials might work in an optical maser. The simplest spectra to understand were those of metals with only a single electron in their outer shell. Townes initially focused on thallium, which he knew from Gould's experiments. However, Schawlow dug into the spectroscopy and found it wouldn't work because of how the atoms shifted between energy levels. They then turned to the soft and very reactive alkali metals, such as potassium, sodium, and cesium, which also had well-known spectra. They settled on potassium for a pragmatic reason—two key wavelengths where it absorbed light fell in the visible part of the spectrum, so Schawlow could study them with a spectrometer left over from his superconductivity research. Their calculations showed that it should work, but they would come to regret the choice. "If I had known how rotten potassium is in some other ways—it's very reactive and easily quenched—I'd probably have picked something else," Schawlow said later.

They also needed a resonant cavity for the waves to bounce around in as they were amplified. Microwave maser cavities were rectangular boxes with holes in the sides for the microwaves to escape. The holes had to be roughly the same size as the waves. That was fine for waves a few centimeters long, but visible wavelengths were shorter than one micrometer—0.001 millimeter. Townes thought of a rectangular box, with holes big enough to let energy in to excite the emitting material and to let some stimulated emission escape. They struggled to understand how to control the light so it didn't bounce off in a million different directions. Finally a simpler idea hit Schawlow either while he was shaving or while he was riding the train back to New York from visiting relatives in Toronto.

Schawlow proposed throwing away the sides of the box, leaving only two reflective ends. For his dissertation research at the University of Toronto, he had used an instrument called a Fabry–Perot, in which light bounced back and forth between two flat glass plates aligned so their reflective surfaces were parallel. In a Fabry–Perot, the plates are big and close together. For an optical maser, he envisioned much smaller plates that were much further apart, on the opposite ends of a long cylinder. As long as the mirrors were many times wider than the wavelength of light, and properly aligned with each other, they would reflect the light back and forth. Better yet, they would control how the light waves resonated within the cavity.

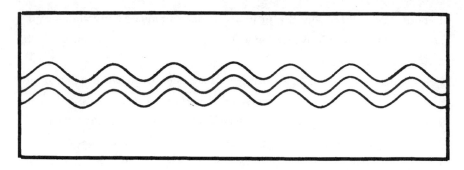

FIGURE 3.3. A Fabry-Perot cavity resonates at wavelengths where an exact number of waves fit into a cavity. (From Jeff Hecht, *Understanding Lasers,* © 1994 IEEE. This material is used by permission of John Wiley & Sons Inc.)

Waves resonate if they fit exactly inside a cavity, as shown in figure 3.3. In that case, the peaks and the valleys of the reflected waves line up with the original wave, reinforcing it. A shower stall that echoes when you sing a low note is resonating at the frequency of the echo. The same happens for electromagnetic waves. Microwaves are long enough that it was easy to make a cavity with dimensions comparable to the wavelength. In fact, the cavity dimensions selected the wavelengths, because only waves that fit exactly could resonate. Other frequencies faded away. This selectivity was a big advantage of the microwave maser, because it selected only the extremely narrow range of wavelengths that fit exactly into the cavity.

An optical maser was a different matter, because it had to be tens of thousands to over a million wavelengths long. At first glance, a cavity thousands of wavelengths on a side would seem to have countless resonances, because the light could bounce around inside it in many different ways. One path might fit 5896 wavelengths, another 5897 wavelengths, a third 5898 wavelengths, and so on, so the cavity would oscillate in many modes, each at its own frequency. Schawlow's idea gave light waves only a single path to follow, bouncing back and forth straight between the two small mirrors. Ideally that would limit the optical maser to oscillating in only one mode, at a single wavelength, making it useful for spectroscopy and communications. (In practice it's a bit more complicated, because stimulated emission is actually at a narrow range of wavelengths, so light can oscillate in a few different modes, but not enough to cause a problem.)

By about February 1958, Schawlow figured the Fabry–Perot resonant cavity would limit light emission to an angle of a few degrees (more than a modern laser). When he heard the idea, Townes quickly realized that the passage of the light back and forth between the mirrors many times would select only the strongest resonant wavelengths. "At that point we felt that we had it," Schawlow said.

The next step was to flesh out the details of their hypothetical optical maser. To make calculations easy, they assumed the mirrors were 10 centimeters apart and each had an area of one square centimeter. Potassium vapor would fill the space between the mirrors. They proposed optically pumping the potassium with violet light at 0.405 micrometers emitted by a lamp also filled with potassium vapor. They hoped that light would excite the potassium ions to a state where they could be stimulated to emit infrared light at one of two wavelengths, 2.71 and 3.14 micrometers. After releasing that energy, the atoms would drop to lower energy levels. It was a three-level maser translated to optical wavelengths. Schawlow and Townes thought they had a chance for stimulated emission at either wavelength, and calculated that under the right circumstances their simple optical maser would oscillate. They didn't expect to get much light out; all they wanted to do was to show that the idea worked.

In February or March of 1958, they decided to publish their theoretical analysis before they tried to build anything. To Townes, it was a matter of establishing priority. He had learned the hard way from the microwave maser that it didn't pay to keep quiet in a fast-moving field. Although he and Gordon had worked out the idea of the microwave maser before they made the first one work, they delayed publishing the proposal until they succeeded in the lab. In the meantime, Prokhorov and Basov had developed the theory on their own and published it. Townes did not intend to be scooped again. He had graduate students whom he could put to work building an optical maser using their potassium-vapor scheme, but they would need time. Schawlow felt he lacked the resources and equipment to charge ahead with the potassium system. They sat down and quietly formulated their ideas. They considered solid-state materials, which were making rapid headway in microwave masers, but didn't come up with a good example of how one would work for light.

Both Townes and Schawlow were working on other projects, so their progress was slow. Only in early summer did they circulate a draft around Bell Labs for comments, standard procedure before Bell scientists submitted papers. Clogston and some skeptical engineers insisted they go back and more carefully examine how light waves would oscillate within the cavity formed by the pair of mirrors. Townes worked out the details, and the two submitted their paper to *Physical Review* in late August.

They also ran the paper past Bell's patent office. The lawyers initially saw nothing worth patenting, saying that an optical maser was irrelevant to the parent American Telephone & Telegraph Corporation because light wasn't used in communications. Townes took strong exception, and by citing potential uses in communications managed to get them to file an application at the end of July. The lawyers took his comments to heart; the first three claims in the patent covered uses of optical masers in communication systems.

Only after the patent was filed and the paper was in the mill at *Physical Review* did Townes start discussing the idea outside Bell Labs. In late August he started sending copies to colleagues around the country. By the time the paper finally appeared in the December 15, 1958 *Physical Review Letters*, many maser researchers had already seen it. Properly dense with detail and containing 35 equations, the paper made an eloquent case to physicists that optical masers might be within reach.

The paper set the research community buzzing. Townes and Schawlow knew they had something good. In a footnote, they conceded that Prokhorov and Dicke had separately proposed making resonators from a pair of parallel plates, but both had thought of putting the mirrors close together. Townes and Schawlow had developed the idea much further. They had good reason for thinking they had laid the groundwork for a promising new field. In a sense, they fired the starting gun for development of what they called the optical maser.

Schawlow and Townes recognized key features that would define a laser, such as a narrowing in the range of emitted wavelengths once it began operating. Yet their paper was far from the final word, or a blueprint for building a practical device. Despite their analysis of modes inside the laser cavity, they wrote little about how the light would emerge from the cavity. They noted that putting highly reflective mirrors on the ends of the cavity would make the light spread out more slowly. If their one-square-centimeter mirrors reflected 98% of the light back into the 10-centimeter cavity, they predicted the light would spread out at an angle of 0.11 degree (0.002 radian). Perhaps because they were used to working with microwaves, they said nothing about the importance of producing a narrow beam or concentrating light to deliver high power onto one spot on a target.

They expected the optical powers to be low, with pump power of around a watt generating about a milliwatt of stimulated emission. They gave no details about the optical arrangement that would be needed for optically exciting potassium vapor or any other laser material.

They also hadn't reckoned with Gordon Gould, who had caught the same intellectual bug. Weeks after talking with Townes, Gould had left Columbia to work on what he called the "laser."

4 THE OUTSIDER'S INVENTION: THE LASER

TOWNES MAY HAVE UNDERESTIMATED Gordon Gould, but Gould harbored no delusions about Townes. Gould had already been thinking about an optical version of the maser, and he knew that Townes would be formidable competition. Yet Townes's interest also validated Gould's intuition that the idea was important. Gould sometimes put off unpleasant tasks, but his conversation with Townes pushed him to action.

The possibility of collaboration did not come up in their discussions. It was probably just as well; they would not have made a good match. Both tended to guard their own ideas, but for different reasons. Townes had a natural tendency to be reserved. Gould distrusted what he saw as a hostile establishment. The two men had strikingly different goals and intellectual styles. Townes relished the intellectual challenge of physics and thrived on recognition by his peers. He approached problems deliberately and systematically. Gould thought intuitively, dreamed of being an independent inventor, idealized Thomas Edison, and hoped to get rich. It wouldn't have worked.

Their personal styles were strikingly different as well. Townes was an upright pillar of the establishment. He worked long hours six days a week, but saved Sunday for church and family. Gould's work habits were more erratic. Unlike many younger doctoral students, he had a life outside the university, and sometimes his lab would appear vacant for days. He liked to sail and had a taste for the good life that grad students rarely had a chance to enjoy.

Their ancestral anecdotes highlighted their different attitudes. Townes took a certain pride in being a Mayflower descendant on his mother's side. Gould could trace his father's family back to Puritan New England, but he preferred to cite the French pirate Wonny La Rue on his mother's side, and named a sailboat after him. It was a symbol of Gould's anti-establishment attitude. His father, Kenneth Miller Gould, was the long-time editor of *Scholastic* magazine, and the author of a book on scapegoats in history. His mother was active in Democratic politics, and Gould embraced the family's liberal idealism as he grew up in the 1920s and 1930s in Pittsburgh and the affluent New York suburb of Scarsdale. Born in Manhattan in 1920, just five years after Townes was born in South Carolina, Gould grew up in a different world.

Gould's mother started him building things when he was only three, and he did well in school. After graduating from Union College, Gould began graduate school at Yale, studying optics and spectroscopy, but World War II intervened, and he wound up working on the Manhattan Project. He got involved with a woman coworker who had a penchant for Marxism, and Army Intelligence got them both sacked—along with other suspected leftists—in early 1945. The experience pushed Gould's politics further to the left, and his girlfriend pulled him into Communist circles. They married in 1947, but the 1948 Soviet takeover of Czechoslovakia disillusioned the idealist in Gould with the reality of Communism, straining their marriage.

Gould held a series of jobs and tried to make a living as an inventor before landing a job as a physics instructor at City College of New York. In 1949, he enrolled in Columbia's physics program, hoping that what he learned would make him a better inventor. By the time he finished his course work two years later, he had split from his first wife and her communist friends. Kusch put him to work helping Peter Franken finish his thesis research with an atomic beam system, so Gould could learn how to use the complex machine for his thallium research. Gould continued teaching to support himself, but that left little time for his thesis research until the summer of 1954. Then a wave of anti-Communist hysteria swept through New York, and the New York City Board of Higher Education demanded Gould tell them who had been in his Marxist study group. Gould had left the group, but refused to name the people, and was fired. The dejected Gould was ready to drop out, but Kusch was furious at the witch hunting and would have none of it. To make ends meet, Gould moved in with Ruth Hill, a biophysicist he had been dating, an open cohabitation that was unusual at the time. She had a Columbia Ph.D. herself, and agreed to support Gould so he could finish his. They married in 1955, and Kusch eventually found a part-time job for Gould so he didn't have to depend totally on his wife.

When Rabi introduced Gould to optical pumping, Gould saw a new range of possibilities. The immediate one was exciting thallium atoms so he could measure the energy of transitions in the radio-frequency range to complete his dissertation research. His hand-me-down apparatus was primitive even for the time, and other techniques hadn't been adequate. Optical pumping worked well, and soon he had the data he needed. That led Gould to realize that optical pumping might be able to excite microwave emission, by producing the inverted population needed to make a microwave maser. He talked with Townes about this in 1956, and Townes asked him to describe it at a weekly Columbia physics seminar. Gould also asked Townes how to file for patents, and Townes described how he had done it.

Gould took the next logical step, and began to wonder if optical pumping could also excite light emission. When an atom absorbed light, it gained the amount of energy contained in the light. The energy released as a microwave was only a tiny fraction of the energy absorbed as light. It was as if the light raised the atom to the top of a ladder, but microwave emission dropped the atom down only one step. Gould could see the extra energy sitting there, waiting to be released. He knew that masers could stimulate microwave emission. Perhaps the same principle could work for light, dropping the atom further down the energy-level ladder.

The idea distracted Gould from settling down to write his dissertation. The research had dragged on for years, and after he thought he was done, Kusch had asked him to make more measurements that kept him busy months longer, through the end of the summer in 1957. After Townes asked him about thallium lamps, Gould brooded over the ideas he had discussed with Townes. Their second conversation made him realize he needed to write up his ideas as the first step in filing for a patent.

Gould was a careful man, who moved and spoke slowly and seemed laid back. It was easy to underestimate him among the more conventional stars in the pressure cooker of Columbia's physics department. Gould had problems finishing projects, and his long stay at Columbia gave some the impression that he was not a hard worker. Yet once motivated, Gould could focus on a problem with tight intensity. The realization that he had to compete with Townes drove him. Gould holed up in his apartment study until late at night, smoking heavily, checking references, and working out his ideas on paper. He kept it up night after night for several days. The critical inspiration hit him as he lay awake in bed on Saturday night. If he put a pair of mirrors on opposite ends of a long tube, aligned so their surfaces were parallel, light would bounce back and forth between them and through the tube. Put something inside the tube and invert the energy level populations, and light passing through the tube could stimulate emission (Fig. 4.1).

Spontaneous emission of a single photon by one atom in the tube can trigger a cascade of emission, but only if it is in the right direction. Stimulated emission is in the same direction as the spontaneous wave that stimulates it, as well as at the same wavelength. If the spontaneously emitted photon heads toward the sides of the tube, it leaks out, along with any photons it stimulates on the way. That way the light disperses with little amplification. Gould realized that something else would happen if the original spontaneous emission was aligned so the light would bounce back and forth between the two mirrors: as long as the mirrors had their flat surfaces precisely parallel to each other, and exactly perpendicular to the axis between them, the light would bounce back and forth between them indefinitely. On each pass it would stimulate more emission, increasing the power level. Any light that wasn't aligned exactly along that axis would leak out

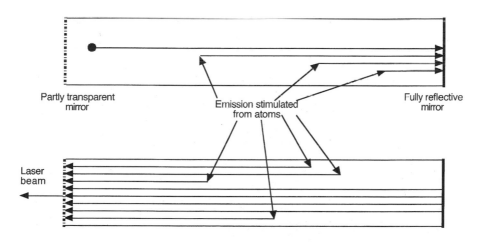

FIGURE 4.1. Spontaneous emission from one atom (black dot) stimulates others to emit more photons, which bounce back and forth between mirrors at the ends of the cavity. The laser beam emerges from a partly transparent mirror.

eventually, but the intensity of the light bouncing back and forth would increase until it reached some limiting value.

That limit depends on the nature of the stimulated emission medium, losses within the cavity, and the fraction of the light transmitted through the output mirror. No one had worked out those details yet, so the potential power was unclear. Gould initially estimated the output would be a power in the range of watts—the output of a bright light bulb, not of a blast furnace. (The watt ratings of light bulbs are for electrical power consumed; only a fraction of that electrical power is converted into visible light.) But Gould knew optics well, and could see that his scheme was more than a bright light bulb. Bouncing the light back and forth between the two mirrors would align the light into a tightly focused beam, and some of that light could emerge if one mirror was partly transparent. He realized that the raw power was secondary to the ability to concentrate the light. "What was important was being able to focus that power into a very tiny spot, and thereby creating an intensity which was unprecedented," Gould said.

Light from a bulb spreads out in all directions; stand a foot away and you can't feel its warmth. But by reflecting light back and forth through a medium that amplified light, the pair of mirrors would form a beam of parallel light that could—at least in theory—be focused to a spot about a wavelength across. Gould estimated this concentration of power could produce temperatures as high as the surface of the sun, and perhaps high enough to compress atomic nuclei to the point where they fused together. Although the power would be concentrated onto a tiny spot, the intensity in that spot would be enough to burn holes in hard materials.

Every kid interested in optics has focused sunlight through a magnifying glass onto a blindingly bright spot that can burn a hole through paper. Gordon Gould had gone a step beyond. He had invented a beam of directed energy—the laser beam. The beam vastly amplified the importance of stimulated emission. "I understood that the amplification process would work, right from the beginning, early in 1957, but it was only when I realized that you should shape it into a beam that I realized how useful it could be," Gould recalled.

His wife Ruth found Gould up and working intensely on his patent ideas Sunday morning. She wanted him to concentrate on writing his doctoral dissertation. She had earned a Ph.D. in biophysics before their marriage and thought it an essential academic credential. Gould insisted that patenting the laser was more important. He had a tiger by the tail, and he knew Charles Townes wasn't going to give him time to finish his dissertation. His wife tried to get him to go out into the nice Sunday afternoon, but he stayed inside, smoking and working on his patent notes. Gould worked well into the next week, poring through books, making calculations with a slide rule and by hand on large sheets of paper, and thinking about the results. Then he organized his thoughts and compiled the results, writing in fountain pen in a bound notebook, in the formal style of a laboratory notebook that could document his claims in a patent case. He had learned the process from Townes. At the top of the first page he gave his own name to his idea: "Some rough calculations on the feasibility of a LASER: Light Amplification by the Stimulated Emission of Radiation" (fig. 4.2).

The maser was Townes's baby. His subsequent use of the term "optical maser" was a reminder that it was derived from the maser, and—secondarily—that it was an outgrowth of Townes's earlier work. In a sense, Gould merely adapted the maser acronym, replacing the M from "microwave" with an L from "laser." Yet the change stressed that his idea was fundamentally different, not just a derivative.

The nine pages of handwritten notes that emerged from Gould's intense days and nights of effort were more than a first draft. Like the final version of a school paper before the computer era, they were neatly written and fairly polished, with only a few corrections. Their legal purpose was to show what he knew and when he did the work. At Columbia or Bell Labs, physicists normally had another scientist read their notes, then sign them with a date, noting they were "read and understood." Schawlow had done that for Townes on the microwave maser. But Gould didn't want to take his notebook to Columbia, where the competition worked. Instead, on November 13, 1957, he walked to a nearby candy store, run by a man named Jack Gould, who was no relation but was a notary public. The inventor showed the notary the notebook, and asked him to sign, date, and stamp all the pages. The notary obliged, squeezing his stamp and signature sideways on the crowded pages. The inventor added his own signature, and bought a carton of cigarettes to take home as he paid the notary.

Although Gould had worked for years with thallium, his laser scheme was based on potassium, the same element that Schawlow and Townes selected. To Gould, it was a logical and practical choice. Thallium had to be heated to 1500°C before it vaporized. Thallium was also toxic, and Gould had found that it settled out on cooler surfaces and gummed up the works of his thesis experiment. He was probably sick of the stuff. Potassium seemed much simpler. Unlike thallium,

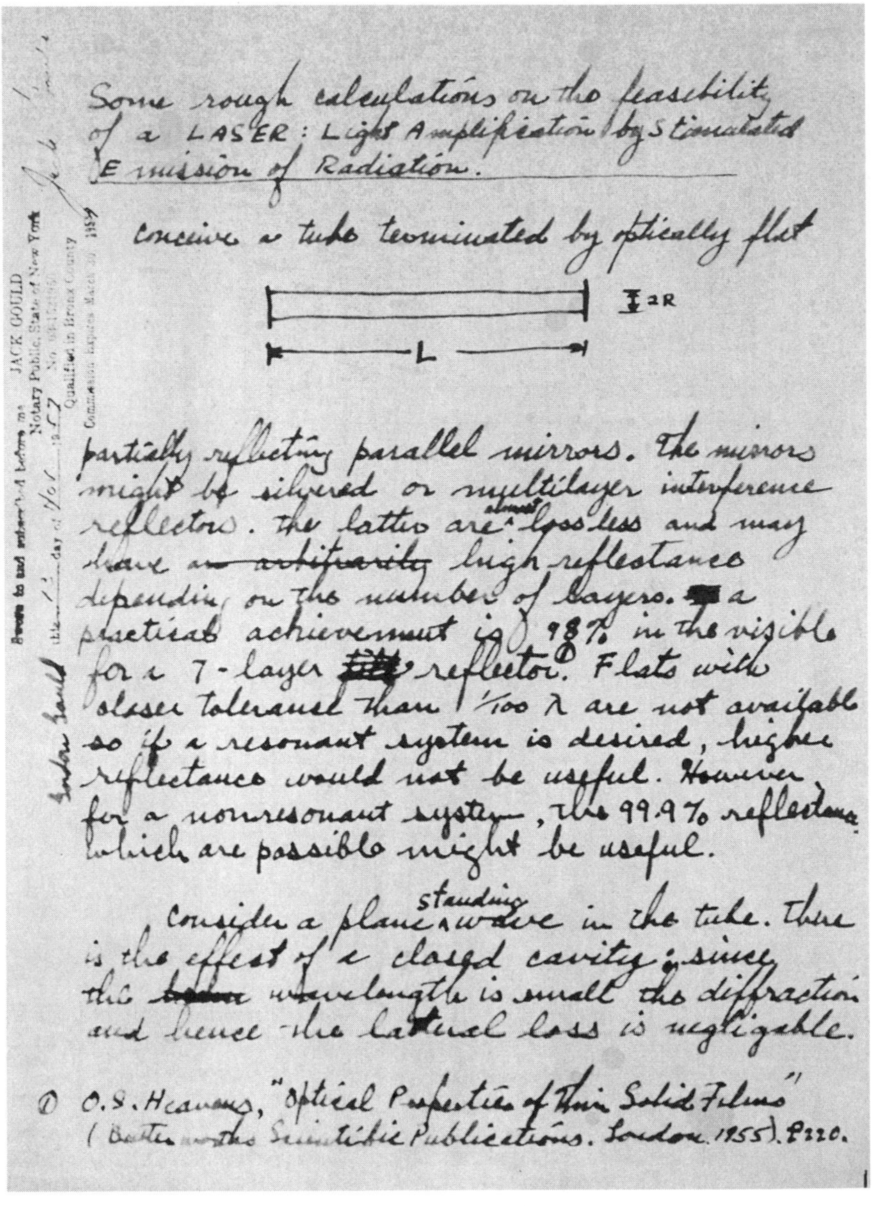

Some rough calculations on the feasibility of a LASER: Light Amplification by Stimulated Emission of Radiation.

Conceive a tube terminated by optically flat

partially reflecting parallel mirrors. The mirrors might be silvered or multilayer interference reflectors. The latter are almost lossless and may have ~~an arbitrarily~~ high reflectance depending on the number of layers. A practical achievement is 98% in the visible for a 7-layer reflector.① Flats with closer tolerance than 1/100 λ are not available so if a resonant system is desired, higher reflectance would not be useful. However for a nonresonant system, the 99.9% reflectance which are possible might be useful.

Consider a plane standing wave in the tube. There is the effect of a closed cavity; since the ~~beam~~ wavelength is small the diffraction and hence the lateral loss is negligible.

① O.S. Heavens, "Optical Properties of Thin Solid Films" (Butterworths Scientific Publications. London. 1955). P.220.

Sworn to and subscribed before me

JACK GOULD

Notary Public, State of New York

Qualified in Bronx County

Gordon Gould

FIGURE 4.2. Gordon Gould's description of the laser in his first notebook. (Courtesy of Gordon Gould)

it has only a single electron in its outermost shell, which it readily gives up to react with other materials. Potassium is so reactive that it burns in water, and reacts with many other materials, but it vaporizes at 760°C and isn't toxic. Its optical properties were well-known, and Gould saw it as a reasonable starting point to demonstrate his concepts.

Gould envisioned filling a transparent tube with potassium vapor, and shining a bright light through the walls to excite the potassium atoms. His initial vision of the laser cavity had flat mirrors on the two ends, like what Schawlow suggested to Townes.

All in all, the basic physics of Gould's laser was very similar to that of the Schawlow–Townes "optical maser." His first notes focused specifically on potassium vapor; the example that Schawlow and Townes considered in greatest details. Yet Gould's intellectual approach to the laser differed profoundly from theirs. Gould's style was intuitive rather than analytical. An intuitive physicist looks at a system and visualizes how it works, taking it apart and putting it together in his mind and in the lab. In contrast, Townes always took a more systematic and rigorous approach, starting with theoretical principles and analyzing the possibilities with mathematical formulas. The difference is subtle but important. Gould started from visualizing how a system would work, then sat down and calculated. Townes started with formal analysis, then built up a sense of how the system worked. Gould was instinctively practical; Townes started from fundamentals. Both approaches work, but they yield different viewpoints. Schawlow fell between the two.

A second key difference was experience. Gould came from an optical background and worked with microwaves at Columbia. He had worked in industry and tried to earn a living by inventing things that met specific needs. Schawlow had some experience with optics, but he and Townes had worked mostly with microwaves. Their experience was largely in research. Schawlow and Townes described their optical maser as an oscillator, generating a resonant wavelength. Their paper devoted considerable attention to modes of oscillation within the resonator, and they were particularly interested in confining oscillation to one mode so the light would be emitted at a single wavelength. They expected the device to emit a beam, but they didn't expect it to be powerful; all they wanted was a research tool. Gould's optical view showed him that oscillating back and forth between the pair of mirrors would concentrate the light energy in a highly directional beam. His intuition leaped ahead to recognize that a laser might generate significant power, and deliver it in a beam of concentrated energy. The bright light of invention stared him in the face, and the compelling vision of concentrating energy in a powerful beam seized Gordon Gould.

Ruth Gould still wanted her husband to finish his dissertation. When she saw his laser obsession, she suggested he try to do his dissertation on the laser. Gould shrugged her advice off, convinced that Kusch was too much of a pure scientist to let him work on anything so practical. Townes might have been more receptive, but that would have meant that Gould had to start a new round of thesis research, and would have to share both credit and patent royalties with Townes. Nor did Gould seem comfortable working with the establishment, despite the support Kusch had given him after he lost his teaching job. Finishing his dissertation would lead him to a job in industry or a university, but Gould was not an organization man. His dream was being an independent inventor, not working for somebody else.

The more Gould thought about the laser, the more obsessed he became with the idea. The government had persecuted him for his politics, but he was an idealistic dreamer rather than a revolutionary. Despite his sympathy for the downtrodden, Gould didn't want to live a life of poverty. He had enjoyed sailing since his teens, and kept a small sailboat at his brother's house. He liked the idea of getting rich, and the laser looked like it could be his chance. Gould knew he was in the right place at the right time. He realized the laser "was going to be the most important thing I ever got involved with in my life." It was time to make a big, risky bet.

His first step was consulting a patent lawyer. Gould's parents knew a top lawyer at Darby and Darby, a New York patent law firm, which helped get him in the door. In January 1958, Gould visited Robert Keegan, a young staff attorney. Unlike the senior lawyers, Keegan had heard of the maser. Gould worried that Columbia would get a share of his patent rights, but Keegan reassured him that the university's standard patent contract didn't apply because it wasn't part of his thesis work.

Unfortunately, Gould misunderstood Keegan's explanation of a crucial point of patent law. A patent must describe an invention in enough detail for a person "skilled in the art"—such as an engineer trained in the field—to replicate or build it. In patent law, this is called "constructive reduction to practice," and it's equivalent to providing a detailed set of instructions. It does not mean that the inventor has to submit a working model. (The only exception is for perpetual motion machines, where the Patent Office insists on a working model to discourage legions of crackpots from applying.) When Gould left Keegan's office, he thought he would need to build a laser in order to earn a patent.

That fundamental misunderstanding would cost Gould dearly in the short term. Had he sat down with Keegan then and there to start writing an application, he likely would have had his laser patent. At that point, Gould didn't have the in-depth analysis needed for publication in Physical Review, but he probably did have enough to satisfy the Patent Office. He had an initial lead over Townes and Schawlow, who were just figuring out how to make a laser cavity. Prompt filing would have given Gould priority for a fundamental patent, although the Patent Office normally took a couple of years to process and issue it. Schawlow and Townes would still have published the first scientific paper, but they would not have had patent priority. Bell Labs could have fought Gould's claims, but they would have had a very hard time voiding all of them. If his patent had been issued in 1960, Gould might have collected a few million dollars in royalties through its expiration in 1977. Yet as long as he lacked the means to build a laser, Gould could have little influence on the course of the race to make a working laser. Gould's misunderstanding thrust him into the thick of the laser race.

Gould knew he lacked the resources to build a laser on his own, but he was determined to go for a patent. Columbia had most of the equipment he would have needed, but as a graduate student the best Gould could have hoped for was playing second fiddle to a professor—if he could find one willing to work with him on such a practical device. Gould knew his idea was good, and he wanted more. He started looking for a more receptive environment, where he could continue his laser work on the side. He found one at a fast-growing, young Manhattan-based company named the Technical Research Group and usually called TRG.

Founded in 1954 by Lawrence Goldmuntz, TRG had thrived doing contract research for the Pentagon. Gould had already consulted for the company on an Army Signal Corps contract run by Richard Daly. Gould had shown Daly how to optically pump a beam of cesium atoms, and initially the two men hit it off well. Gould liked the company as well. Like many inventors, Gould was part entrepreneur, developing ideas and running with them for all they were worth. TRG's managers were younger, looser, and more aggressive than the Columbia professors. After a series of interviews, they hired Gould to work for Daly on a project to develop atomic clocks that used atomic beams to generate a precise frequency, which could serve as a time standard.

After landing the job, Gould went back to tell Kusch he was leaving. His excuse was that he was broke, and he halfheartedly told Kusch that he would return to finish his dissertation. In reality, Gould could have lived on his wife's income, but his laser dreams had eclipsed the mundane task of analyzing and writing up his thallium experiments. Later Gould lost the thallium data, which surprised friends because he normally was well organized. Yet Gould had clearly lost his interest long before he lost the data.

Once Gould started working at TRG on March 27, 1958, he devoted lunch hours and evenings to the laser project. He pored through references, hunting for transitions he might be able to use in a laser, and began a more detailed and systematic analysis in a new notebook. He broadened his focus beyond optical pumping to consider exciting gas atoms in a different way, by passing an electric current through them so electrons transferred their energy to the gas. It's the same process used to excite light emission from fluorescent lamps and neon tubes, and it can put much more energy into a gas than optical pumping. He dug into old journals to find out which gases could absorb electrons efficiently, and analyzed their energy levels looking for ways to transfer energy and produce stimulated emission.

Like most technology companies, TRG expected employees to sign a form granting the company patent rights to ideas they created on the job. Gould dodged the paperwork for a while, but eventually it became an problem. When Goldmuntz asked why Gould wouldn't sign the agreement, Gould said he was trying to preserve rights to something he had invented before starting work at TRG. Goldmuntz was sympathetic, and had an interim agreement drawn up to exclude the earlier work, which Gould signed. He also asked Gould to write up his ideas so they could finalize the agreement.

Worried about protecting his rights, Gould mostly kept quiet about his private project as he labored overtime on his laser ideas. Only a few times did his enthusiasm overwhelm his caution. As he was working on an atomic beam experiment in a darkened TRG laboratory, he told Maurice Newstein that he was developing a "revolutionary" idea. He also refined his laser ideas. By late August, Gould had written a 23-page analysis in a second notebook that went far beyond his original idea of optically pumping potassium vapor.

When Gould finally described his laser ideas to Goldmuntz and TRG scientists, "I was at first met with blank stares," he recalled. They couldn't imagine how his proposed laser could generate a beam of coherent light. It took him quite a while to convince them that the laser was theoretically possible, and that his device could actually achieve the conditions a laser needed to operate. Newstein and the theoreticians got the idea first, and eventually Goldmuntz, an experimentalist, caught on.

Once he was persuaded that Gould had a way to generate coherent light, Goldmuntz was all ears. Light is coherent if the waves are the same length, and march along with their peaks and valleys aligned in matching phase, as shown in figure 4.3. When a photon stimulates emission of another photon, the two are coherent with each other, and as the photons bounce back and forth through the cavity stimulated emission generates more photons that are coherent with the first. Radio oscillators generate coherent radio signals, but nobody else was talking seriously about generating coherent light at the time. Gould wove a magically persuasive spell, explaining how a laser would work and what he could do with it. Some of the details remained vague, but Gould had Goldmuntz and the other TRG physicists hooked. Goldmuntz told Gould to turn his ideas into a research proposal to submit to the Pentagon.

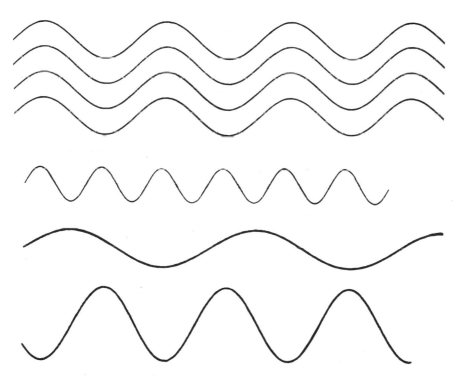

FIGURE 4.3. Coherent and incoherent light. (From Jeff Hecht, *Understanding Lasers*, © 1994 IEEE. This material is used by permission of John Wiley & Sons Inc.)

The physics Gould described was essentially the same as what Schawlow and Townes had worked out. It was natural enough; they had started from similar points toward the same goal. They described similar ideas in different ways; Gould talked of coherence, while Schawlow and Townes merely said the light would be at a single wavelength. For most practical purposes, those were different ways of saying the same thing, yet Gould's choice of words was better at grabbing attention. Electronic oscillators could generate coherent waves at radio and microwave frequencies. Townes and Schawlow, familiar with microwaves, didn't immediately think that coherence was a special feature of optical masers because it wasn't a special feature of microwave masers. Gould worked with light, and he knew that coherent light would be something very special.

The two visions also differed in a much more important aspect. Townes and Schawlow had thought of the optical maser as a low-power device, emitting a trickle of light energy for spectroscopic research or perhaps short-distance communications. Gould had more ambitious ideas. Electric discharges could deposit much more energy in a gas than optical pumping, and he hoped to extract that energy as a laser beam. He "was thinking in terms of incredible power; he visualized the beam, he thought of things like communications from here to Mars," recalls Paul Rabinowitz, who worked with him at TRG. Gould envisioned a powerful beam; one that would not just relay a signal from point to point, but could concentrate enough energy to vaporize a spot of material, cutting or drilling holes.

Gould's laser proposal invoked visions of powerful beam weapons dating back to the heat rays that Martian invaders carried to Earth in H. G. Wells's 1897 novel *The War of The Worlds*. Ray guns, death rays, and blasters had echoed through the pages of pulp science fiction magazines, seared the screens of science-fiction movies, and spread from radio serials to television. Military agencies had asked if such weapons were feasible, but scientists had always told them it was just science fiction.

Yet Gould was not writing science fiction. He was proposing an idea that the physicists could see was reasonable. Bounce a beam of light back and forth through an excited gas between a pair of mirrors. On each pass through the gas, stimulated emission would amplify the light a little. One mirror would transmit a fraction of that light energy, which would become the beam. Another small fraction of energy would be lost in the gas and optics. As long as each pass through the gas added more light than was lost, the power would build up inside the cavity and in the beam outside.

The power doesn't have to increase much each time. Suppose that after you count all the losses and the energy in the beam, the power increases only 1 percent per round-trip through a laser cavity one foot (30 centimeters) long. After the first pass, the light power would be 1.01 times the original input. The second pass it would increase another 1 percent, to 1.02 times the original input. At first the growth would be very slow; after 20 round trips, the power inside the laser would be only 1.22 times the original, but it would grow. After 100 times, the power would be 2.7 times the original, not merely double the result of adding 1 percent a hundred times. After 1000 passes the beam would contain 20,959 times more power. After 10,000 passes, the beam would contain 1.6×10^{43} times more power. And 10,000 round trips back and forth a foot-long cavity would take only 20 millionths of a second.

The power couldn't really increase so dramatically. At some point, stimulated emission would drain the energy from all the excited atoms. Other losses might increase. But the TRG physicists could see that Gould's laser had the potential to produce a powerful, tightly focused beam of light. It was just the sort of thing that should interest the Pentagon.

Once assigned to work on the laser, Gould seemed to slow down a bit—as if the project was more compelling when forbidden. But on December 1 he turned his final draft in to Daly, and on December 16, 1958, TRG mailed a finished 120-page proposal to the Air Force Office of Scientific Research and the Army Signal Corps. The company asked for $300,000, big money in 1958 dollars.

The issue of *Physical Review* containing the Schawlow–Townes paper carried the date December 15, 1958. The laser race was off and running.

5
BELL LABS TAKES THE EARLY LEAD

THE SMART MONEY would have bet on Bell Telephone Laboratories as the laser race started. Townes and Schawlow's work gave Bell a solid head start over everyone but Gould, who was still struggling for support. Bell was in its glory days, considered to be the world's premier industrial research laboratory, and the driving force behind the steady advances of communications technology. In 1956 Bell bagged the Nobel Prize in Physics for the transistor, although by then only one of the three recipients—Walter Brattain—remained on the Bell payroll. Other companies tried to model their own research after Bell's, even borrowing from the architecture of the sprawling suburban campus in Murray Hill, New Jersey.

Money for the labs flowed freely from the parent American Telephone & Telegraph Company, which at the time had a regulated and highly lucrative monopoly on telephone service in most of the United States. (Most Americans called AT&T "*the* telephone company," and didn't know that well over a thousand smaller companies also provided local phone service, mostly in rural areas.) The best talent money could buy was given the chance to pursue a wide range of basic and applied research. Many top-notch young scientists used the labs as a stepping stone to academic careers, as Townes had done in the 1940s. Others stayed for their entire careers, glad to avoid teaching and academic politics.

AT&T's press office heralded the progress made possible by Bell Labs research, but the real goal was fuelling the company's growth. Besides its scientists, Bell Labs had a superb engineering team that designed, built, and tested new telephone equipment that was produced by AT&T's manufacturing arm, Western Electric. The company needed to keep up with the steady growth in the number of telephone lines since the late 1930s, and the increase in both local and long-distance traffic. Hoping to fuel further growth in traffic, Bell began testing Picturephone, an experimental video telephone system, in 1956. Management thought Picturephone would become the next big thing, and began planning expansion of the telephone network. Time would show they were wrong about Picturephone, but right about expanding network capacity.

Bell planned to expand network capacity by following the trend of electronics toward operating at higher and higher frequencies. The higher the frequency of a signal, the more information it can carry. Radio frequencies above 50 kilohertz carry voices for AM radio. Frequencies above 50 megahertz carry television. Microwave frequencies of a few gigahertz can carry many voice or video signals simultaneously, and in the 1950s telephone companies used chains of microwave relay towers to carry many phone calls simultaneously over long distances. To meet future demand, Bell was developing higher-frequency microwave systems to transmit at 60 gigahertz. Air and moisture absorbed that frequency, so it would have to be transmitted through buried pipes called waveguides.

Bell's patent lawyers hadn't been interested in Townes's optical maser proposal because they didn't see where light fit into the picture. However, top technical managers knew that visible light is an electromagnetic wave with frequency around 600,000 gigahertz, so it had the potential to provide far more transmission capacity than microwaves—if the light had a very narrow range of wavelengths. The light sources of the 1950s didn't.

The very first radio signals were essentially bursts of low-frequency noise that transmitted Morse code by turning the noise off and on. Their big advantage was that low-frequency radio waves travel long distances through the atmosphere. The real triumph of radio came later, with the invention of the radio oscillator, which emits a single pure frequency to serve as a carrier wave. Modulate that carrier wave with a signal, like a person's voice, and radio wave can carry the signal to distant points. As long as the carrier is a pure frequency, the receiver can subtract the carrier to recover the original signal, as radio and television do today.

The sun, light bulbs, and other sources of light are like the first radio trans-mitters, generating a wide range of light wavelengths. From the viewpoint of a communications engineer, their light was noise in the same sense as the first radio signals—they covered a broad range of frequencies. The optical maser was different, Schawlow and Townes told Rudolf Kompfner, Bell's associate executive director for research. Like a radio oscillator, it should produce a single pure wave-length or frequency. Kompfner immediately realized that the optical maser might be the basis for a future generation of communication systems with capac-ity far beyond what 60 gigahertz could offer. Kompfner didn't expect optical communications to be ready soon, but over its decades as a telephone monopoly, AT&T had grown accustomed to planning far into the future.

Bell Labs management controlled equipment budgets, but the senior people—called members of the technical staff—had wide latitude in picking their own projects. Kompfner didn't tell the senior staff what they had to do, but he did go around talking up ideas that he thought were promising. People listened because he was well-respected and the rich conceptual seeds he scattered often fell on fer-tile ground. One of these was the laser, although for years Bell officials insisted on calling it the "optical maser" because that term had originated with Townes and Schawlow working on at the labs.

Those already busy with interesting projects continued with them, but Schawlow—who was losing interest in superconductivity—was ready for a change, and Clogston encouraged him and became involved himself. Schawlow wasn't strongly competitive, so he decided to investigate solids, which he and Townes had discussed only briefly. It was a logical choice. Solid-state physics was hot at Bell, which had just abandoned research in gas discharges as a relic of the vacuum-tube era. Working at Bell after the transistor, Schawlow said, "you felt that if you could do anything in gas, you could do it better in a solid."

The transistor had made semiconductors the hottest topic in solid-state physics, but Schawlow had no idea how to make a laser from a semiconductor. Instead, he turned to transparent solids, which like gases can transmit light to atoms that can be excited by optical pumping. He and Townes had had a critical insight about solids while writing their paper: Although solids can absorb light over a broad range of wavelengths, they tend to emit light only in a narrow band. That suggested that optical pumping of a solid might excite atoms to a state where they could be stimulated to emit light in a narrow band for a laser, but it gave few clues how to go about it.

Schawlow started with a crystal readily available at Bell, synthetic ruby, made by adding chromium to aluminum oxide. Ruby was an important type of microwave maser, and Bell scientists had grown crystals containing different amounts of chromium to study their properties. They were happy to lend samples to Schawlow, and didn't care if they got them back.

At first glance, ruby looked quite promising. The chromium atoms in ruby absorb violet, green, and yellow light, then release some of the energy as deep red fluorescence. Normally the chromium atoms emit the fluorescence spontaneously, but Schawlow wondered if enough of the chromium atoms could be excited to produce stimulated emission. To find out, Schawlow had to study how ruby absorbed and emitted light, a complex task that required exacting measurements.

Basic physics texts make energy levels look simple by starting with an isolated hydrogen atom, where a single electron circles a single proton at various distances. The further the orbit is from the proton, the higher the energy level. The energy-level ladder looks very simple; the rungs are widely spaced at the bottom, but grow closer the higher up you go. Quantum mechanical rules define what energy levels are possible, and how far apart they are, giving neat sets of emission lines at regularly spaced wavelengths when the single electron drops from outer energy levels to inner levels, as shown in figure 5.1.

With only a single electron, an isolated hydrogen atom is the most simple atom in the universe. The more electrons that orbit the nucleus, the more ways the electrons can interact with each other in complex ways that change the structures of the energy levels. The degree of complexity depends on the number of electrons and how they are arranged. Potassium is fairly simple by atomic standards because the single electron in its outer shell interacts only weakly with the 18 inner electrons. Atoms with more electrons in their outer shells are more complicated. Interactions between atoms make things even more complicated. In a simple molecule containing two atoms, two sets of electrons interact with each other, and the atoms can rotate and vibrate relative to each other. Yet as long as the atoms or molecules are in a gas, they are far apart from each other and don't interact very much, so their energy level structures are relatively simple.

Solids are much messier. The atoms are so close that they bond to other atoms on every side, multiplying the interactions among electrons. The energy levels of the chromium atoms in ruby depend on how they interact with the aluminum and oxygen atoms around them, as well as how the electrons in the chromium atom interact. Put the chromium atoms into a different crystal, and they behave

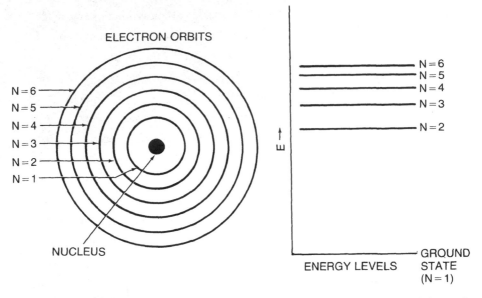

FIGURE 5.1. Energy levels in a hydrogen atom correspond to different orbits of the single electron circling the nucleus. (From Jeff Hecht, *Understanding Lasers*, © 1994 IEEE. This material is used by permission of John Wiley & Sons Inc.)

differently than in ruby. Emerald gets its green color from chromium atoms embedded in a mineral called beryl, a blend of aluminum, beryllium, and silicon oxides.

Deducing the energy levels in ruby was not just a simple matter of plugging numbers into textbook formulas. Schawlow and his technician, George Devlin, had to meticulously measure the wavelengths of light the crystal absorbed and emitted, and match each wavelength to a transition between a pair of energy levels. That required a sensitive instrument called a spectrometer. Schawlow had to borrow other people's instruments until he got approval to buy a sophisticated model with very high resolution, which he said cost more than his house. His boss, Al Clogston, pitched in to help Schawlow with the theoretical analysis, but the going was slow.

Schawlow and Clogston divided their collection of ruby crystals into two groups, depending on their chromium concentration. Crystals containing fewer chromium atoms look pink, while those with larger doses of chromium look red. The difference is more than a matter of color. In pink ruby, the chromium atoms are far enough apart that they have no interaction with each other. In red ruby, the chromium atoms are close enough that they affect the spacing of energy levels. This means pink and red ruby behave differently, as if they are different materials.

Pink ruby fluoresces nicely at two closely spaced wavelengths, 692.8 and 694.3 nanometers, which the human eye sees as red. (Schawlow measured 691.9 and 693.4 nanometers because he cooled the crystals close to absolute zero, which changes the wavelengths.) Yet Schawlow's analysis of chromium energy levels revealed a serious problem. When chromium atoms emit light at those wavelengths, they drop to the lowest possible energy, the ground state, as shown in figure 5.2. That means chromium atoms in the ground state can absorb that light and jump back to the higher-energy state, cutting down the cascade of stimulated emission needed to generate a laser beam. It also means that you have to excite more than half of the chromium atoms out of the ground state and into

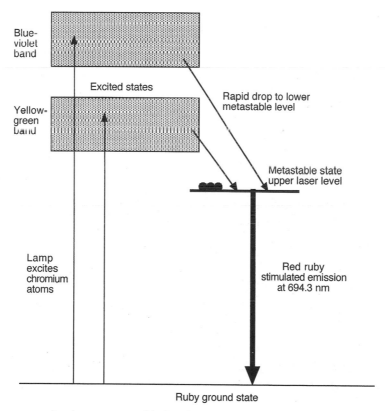

FIGURE 5.2. Pink ruby can emit red light when chromium drops to the ground state, but Schawlow didn't think he could excite enough chromium atoms to produce a population inversion.

the upper state to produce the population inversion needed for stimulated emission to dominate.

That could work if you could excite the chromium atoms very efficiently. Other Bell researchers had studied ruby, and when he asked what they had found, he got bad news. They estimated that pink ruby converted only 1 to 10 percent of the input light into red fluorescence, too low a fraction to do much good. "Without doing any calculations, this really seemed like too much of an obstacle to overcome," Schawlow wrote later.

His study of red ruby held out a faint glimmer of hope. The interaction between closely spaced chromium atoms in red ruby shifts energy levels slightly, so the crystal can glow at a pair of slightly longer wavelengths, 700.9 and 704.1 nanometers at low temperatures. The difference is invisible to the human eye, but critically, the chromium atoms do not drop all the way to the ground state when they emit those wavelengths. They stop at a level just above the ground state. That held out hope that he wouldn't have to excite more than half the chromium atoms out of the ground state in order to get the population inversion he needed. But fellow Bell physicist Stanley Geschwind threw cold water on the idea, saying that excess chromium atoms would absorb too much of the light, and for a while Schawlow abandoned hope. Later Schawlow and Clogston found that reducing chromium concentrations somewhat would alleviate the absorption problem, but because the lower energy level was very close to the ground state, it still required cooling the crystal close to absolute zero.

If that wasn't trouble enough, Schawlow also found that the optical quality of his ruby crystals was poor. Like old glass, the crystals had internal flaws that bent the light unevenly, and that posed a big problem. The logical way to make a ruby laser was to polish the two ends of a ruby rod so they were smooth and parallel, then coat them with shiny films to reflect the light back and forth through the rod. Imperfections in the rod bent light passing through the rod so it didn't reach the opposite mirror, stopping the laser from working. Such difficulties didn't surprise Schawlow, who later said "I had the feeling that, since lasers had never been made, it must be terribly difficult to make one." Frustrated by ruby's limitations, he started studying the properties of other solids, hoping to find one suitable for a laser, but it was slow going.

IN THE SUMMER OF 1958, another young Columbia research fellow followed Schawlow's footsteps to Bell Labs. Born in Tehran in 1928, Ali Javan was the son of an upper-class lawyer who wrote several books, some on human rights. His family had come from Azerbaijan, but they had developed contacts in the Iranian establishment. Mathematics attracted him early; he tinkered with gadgets as a child, and physics became his passion. He came to Columbia in 1949, where he earned his doctorate under Townes, and remained four more years as a research fellow. He became interested in masers while Townes was on sabbatical overseas, and developed (but did not publish) a scheme for a three-level maser independently of Bloembergen and the Russians.

Bright, outgoing, ambitious and full of energy, Javan had a wide-ranging curiosity and seemed constantly in motion at Columbia. Townes urged him to study high-frequency microwave masers, but Javan was also thinking about optical versions of the maser before Bell lured him to New Jersey for an interview. Javan wanted a university professorship, but Bell managers persuaded him the labs could be a valuable stepping stone. On his tour through the labs, he met Schawlow, who told him how a pair of mirrors might act as a laser resonator. The idea caught Javan's imagination, and he rushed back to Columbia that afternoon to start investigating it.

Javan focused on a key problem in developing a laser — exciting the laser medium to produce a population inversion. He wanted to work with gases, because he felt comfortable with them. However, it isn't easy to transfer energy into a gas efficiently. The atoms are spread too far apart, so the energy is likely to leak right through, like light passes through the air. That was precisely the problem he saw with optical pumping; most light passes right through the gas, so little is absorbed. It's useful in the laboratory, but its efficiency is discouragingly low for a practical device. To make matters worse, the special lamps used for optical pumping converted little electrical energy into light at the desired wavelengths. Schawlow and Townes had proposed optical pumping as a way to generate the wavelengths needed to excite atoms up specific steps of the energy-level ladder. However Javan didn't think it would work well enough for a practical laser.

Seeking better efficiency, he decided to try an idea Schawlow and Townes had not proposed but Gould was considering at TRG—passing an electric current through the gas. As current passes through the gas, the electrons hit atoms and molecules, exciting them, and the excited atoms and molecules then hit others, transferring the energy. Javan knew the process could generate light efficiently in neon and fluorescent lamps. In neon lamps, the electrons excite neon atoms which emit red light when they drop back down the energy-level ladder. In fluorescent lamps, the electrons excite mercury vapor, which emits ultraviolet light that produces visible light from phosphors on the inside of the tube.

When he arrived at Bell in August 1958, Javan bought a powerful and expensive magnet typically used to affect the energy levels of atoms in solids. Bell had largely abandoned research on electric discharges in gases, and his managers expected him to study solid-state physics, but Javan left the magnet to collect dust in his lab. Instead, he pored through decades of research on light emitted when an electric current was discharged into various gases. The topic had been hot in the first half of the twentieth century. Physicists had zapped different gases with electrical currents and tabulated what wavelengths they emitted, producing mountains of data. Their main goal was to collect the information they needed to deduce the energy-level structures of the materials they were studying. That was just the information Javan needed to identify gases that might work in a laser excited by an electric discharge.

Javan realized that the spacing and nature of the steps on the energy-level ladder would determine how well a potential laser could work. He wanted to excite atoms to a state where they could remain long enough so that light could stimulate them to emit some of their excess energy and drop to a lower energy level. That required figuring out the complex interactions among the different energy levels in an atom.

It's easy for atoms to make transitions between some pairs of states, and difficult or impossible for them to go between others. Normally atoms drop very quickly from a high-energy state to one with lower energy, as if the rungs on the energy-level ladder were very narrow and greased with something slippery. However, atoms can linger much longer in certain high-energy states, called "metastable" levels, which act like wide steps on the ladder where atoms can rest easily without slipping. Quantum-mechanical rules also prevent electrons from jumping directly between some pairs of energy states because their properties would have to change in impossible ways. That means that atoms can't go directly from some steps on the energy-level ladder to other steps. It's like trying to climb a rickety wooden ladder with some rungs that are broken or rotten, and others that you can't reach from certain angles.

Javan needed an efficient way to pump atoms into the metastable upper laser level and keep them there as long as the laser was working. That meant creating a continual flow of energy from electrons to the excited atoms. Think of it as an assembly line where the electrons lift the atoms from one moving belt to a higher one, where the atoms stay until they are stimulated to jump off at the end of the line. To keep the system running, you always need more electrons to continue raising more atoms to the upper level.

After digging into the literature, Javan picked out a few possibilities. One was firing a beam of electrons directly into pure helium gas in hope of producing a population inversion. The energy levels looked possible, and excited helium was known to emit yellow-gold light, but many important factors remained unknown.

An alternative choice was to mix a bit of neon with a lot of helium. Helium is good at absorbing energy from electrons, and the higher the concentration of helium, the more energy it absorbs. Two helium energy levels that are typically excited by electrons also happen to have nearly the same energies as excited states of neon atoms. The few neon atoms were bound to run into excited helium atoms, and the helium atoms could transfer their energy to the neon atoms by a process that physicists call "collisions of the second kind." (See Fig. 5.3.) The excited neon states are metastable, so neon atoms would stay in them long enough for stimulated emission to extract their energy. Neon is well-known for emitting red light, but Javan's analysis indicated its strongest emission lines were at invisible infrared wavelengths, which he would need special instruments to detect.

The physics looked straightforward to Javan, but he was venturing into new territory. Like Schawlow, he expected the laser would be difficult to demonstrate. Before he charged ahead, Javan wanted to pin down important details, and measure many unknown quantities. He set a series of milestones to convince himself and his managers that he was making progress. A crucial one was to measure the amplification of light in a mixture of excited helium and neon before he went ahead and tried to make a laser oscillator. His calculations indicated that stimulated emission would be weak, amplifying the light only a little bit and making the measurement tricky. Yet it would give him solid evidence of progress in case—as he also expected—it proved hard to align mirrors at the opposite ends of a long tube accurately enough that light would oscillate back and forth between them.

The task looked daunting. When preprints of the Schawlow–Townes papers came out in the fall of 1958, Javan called to enlist the help of another former Columbia student. William R. Bennett had earned his doctorate in 1957 under Chien Shiung Wu, the only woman on Columbia's physics faculty. She was an expert on particle physics, but Bennett's dissertation had been on collisions of the second kind in gases similar to helium and neon. Bennett was a good experimental physicist, and had much more experience working with gas discharges.

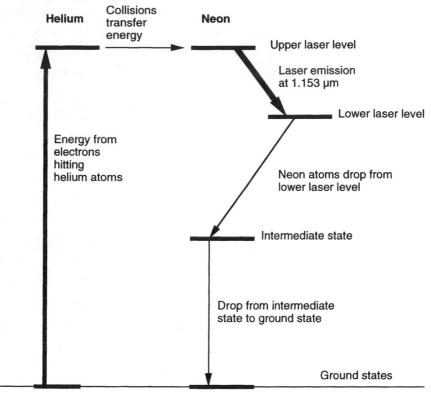

Helium **Collisions transfer energy** **Neon**

Upper laser level

Laser emission at 1.153 μm

Lower laser level

Energy from electrons hitting helium atoms

Neon atoms drop from lower laser level

Intermediate state

Drop from intermediate state to ground state

Ground states

FIGURE 5.3. Collisions of the second kind transfer energy from electrons to helium atoms and then to neon atoms in the helium-neon laser.

Javan suggested they team-up to try to develop an optical maser in a gas discharge using either pure neon and a mixture of helium and neon. Bennett liked both ideas, and after a series of calls they began talking about getting Bell to hire Bennett away from his junior faculty post at Yale. However, that took time to arrange. Bennett missed an interview in January 1959 after catching chicken pox from his children, and the process dragged on until spring before Bell offered him a job starting in the summer.

In the meantime, Al Clogston helped Javan enlist the help of Donald R. Herriott, one of the few optics experts on Bell's technical staff. Optics had earned a reputation as a sleepy field, and Herriott didn't even have a college degree—he joked that his initials D. R. may have fooled the personnel department. However, optics looked like it was going to be a critical technology in making a laser. With light amplification expected to be weak in the gas discharge, Bell would have to push the state of the art to get mirrors that reflected virtually all of the light that hit them. Herriott had learned about such mirrors while working at Bausch and Lomb, then one of America's biggest optical companies, before coming to Bell in 1956, and his practical hands-on knowledge more than made up for his lack of paper credentials.

AS JAVAN GEARED UP HIS RESEARCH at Bell in the fall of 1958, Kompfner invited an old friend from Oxford University to spend his sabbatical at Bell Labs. Knowing that John Sanders had just finished a project in which he precisely measured various physical constants and was looking for new ideas, Kompfner sent him a preprint of the Schawlow–Townes paper in the hope that it would tickle his interest.

Sanders took the bait, but he could only spend January to September of 1959 at Bell. That didn't leave time to tackle a complex project, so he picked a scheme he thought looked simple, exciting pure helium with an electric discharge. Instead of trying a detailed analysis of the whole system like Javan, Sanders started firing electric currents through a tube containing helium in a series of experiments to see what would happen.

Rather than trying to measure low levels of amplification, Sanders put the tube between a pair of flat mirrors and looked for evidence that stimulated emission was triggering laser oscillation. The mirrors had to be aligned extremely precisely so they reflected light back and forth through the tube, but Sanders didn't have sophisticated equipment to align them. He had to settle for putting the mirrors only a few inches apart, leaving little room for stimulated emission to build up in the discharge. His only chance of making that arrangement work was if he could get very strong stimulated emission from the excited helium. He knew it was a long shot, but it was all he could do with his time constraints.

TOWNES WAITED TO LAUNCH Columbia's effort until the fall of 1958, after he and Schawlow began circulating preprints of their proposal and he landed funding from the Air Force Office of Scientific Research. He thought that trying to demonstrate optical amplification by optical pumping of potassium vapor would make a good dissertation project for a graduate student, as the microwave maser had been for Gordon. It wasn't too far from earlier dissertation projects on optical pumping, including Gould's abandoned work on thallium. Columbia had experience with both optical pumping and potassium, as well as the special equipment needed to handle the highly reactive metal.

He first involved an experienced graduate student, Herman Cummins. In February 1959, he recruited a second student, Isaac Abella, who had done optical research at the University of Toronto. Abella originally wanted to work on microwave masers, but Townes handed him a copy of the Schawlow–Townes laser paper and used his professorial power to persuade him to work on the new project. Initially it looked promising. The physical principles seemed straightforward to the students, and they would work directly with Townes, who had analyzed the system in detail.

They soon found the devil was in the details. Optical pumping required a transparent tube, but potassium vapor reacted with the hot glass, darkening it so little light reached the metal vapor inside. Substituting sapphire for the glass tubes reduced potassium absorption but caused other problems. Their experiments required purer potassium vapor than they could buy, so they had to distill the metal to purify it. Heating the vapor made it even more reactive, leading to some minor explosions which slowed progress but fortunately caused no disasters.

They even had problems with the geometry of the experiment. Cummins initially surrounded the tube containing the potassium vapor with four long, thin, potassium lamps, but that didn't transfer the light efficiently. Townes suggested placing the tube containing the potassium vapor and a long potassium vapor lamp inside a reflective elliptical cylinder. He relied on an ingenious trick based on the geometry of an ellipse: Draw rays outward from one focus of an ellipse and have them bounce off the ellipse itself, and they will meet at the other focus. Townes realized the same effect would focus light from a lamp at one focus onto a tube at the other. If the lamp and the tube were aligned along the axis of an elliptical cylinder with a reflective surface, the mirror would focus light from the lamp onto the whole length of the tube. It was an elegant idea, but it couldn't overcome the weakness of their potassium lamp.

6

STIMULATING THE EMISSION OF MONEY

TRG'S LIMITED RESOURCES kept Gould a few steps behind Bell Labs. TRG was a small military contractor; it didn't have piles of money from one of the world's largest corporations to invest in research. TRG's research money came from military agencies, and finding the right agency to foot the bill was always a challenge.

Goldmuntz began 1959 trying to hustle research money from two agencies that had funded earlier TRG projects, the Air Force Office of Scientific Research in Ohio and the Army Signal Corps in Fort Monmouth, New Jersey. Gould came along, waxing eloquent about the laser's promise, but the two failed to convince either Army or Air Force scientists. Aerojet General, a large California aerospace company that owned 18 percent of TRG, also passed.

The problem was the nature of the project. Gould and Goldmuntz were convinced the idea was good, and the potential payoff was huge. However, the proposal also carried a real risk of failure. No one had ever built any kind of laser, much less the powerful types with potential military uses. The military research establishment didn't like risk; they were cautious types who liked to take small steps with predictable outcomes. That left a crucial gap in military research—no one was willing to bet on long-shot ideas with potentially huge rewards.

The Pentagon had come face to face with that gap a year earlier when they found the Soviet Sputnik satellite orbiting the Earth. To fill it, the government chartered the Advanced Research Projects Agency to support risky projects that might give America a vital edge in the long term, as well as prevent future unpleasant surprises. ARPA's job was to nourish crazy ideas that might just be feasible, high-risk projects with a high potential payoff. The agency didn't have to meet near-term needs or worry about being taken to task if some of its projects didn't succeed. They were supposed to push the boundaries of possibility, and they wouldn't be doing their job if everything succeeded. Goldmuntz decided ARPA was the right place for the laser project. He trimmed Gould's list of possible civilian applications for lasers, and concentrated on potential military benefits.

Contacts are an important part of the defense contracting business, and Goldmuntz had one at ARPA, a physicist named Richard Holbrooke. He wasn't an expert in optics or microwave masers, so he asked for help from Paul Adams, a patent attorney with a solid technical background who also worked for ARPA. Adams had a reputation for being brash, loud, and offensive, Holbrooke warned Goldmuntz. He lived up to that reputation as he walked into the ARPA meeting room, saying, "These guys know how to make coherent light? They'll have all the secretaries in this office menstruating in phase."

Yet once he settled down, Adams listened attentively as Gould pitched his idea. Many of the laser's attractions came from fact that optical wavelengths are tens of thousands of times shorter than microwaves. A phenomenon called diffraction causes waves to spread out at an angle that depends on the wavelength divided by the size of the source. The smaller the wavelength is, compared to the source, the smaller the angle. A typical microwave dish is no more than a few wavelengths across, so it spread out at ten degrees or more. Light waves are shorter than 0.001 millimeter, so if they emerged from a tiny one-millimeter source, they would spread at only about a 0.05 degree angle. That meant that, in principle, a laser could project bright spots onto distant objects, or even put a visible spot on the moon. That opened the possibility of long-distance communications through air or space in a beam that could be tightly focused to avoid enemy eavesdroppers.

Gould also suggested that laser beams could probe objects and sense things at a distance. He proposed laser equivalents of microwave radars, which could reveal much finer details because optical wavelengths are much shorter.

In one sense, those ideas were not radically new. Lighthouses had long generated bright beams from whale-oil lamps and electric arcs, focused them with giant lenses and curved mirrors, and swept them across the sea to alert sailors. Searchlights did the same thing. Gould's radically new idea was generating coherent light, with the waves marching neatly in phase, which behaved much better than ordinary "incoherent" light because the waves were better organized. A coherent beam would inevitably spread a little bit with distance, but nowhere near as much as the incoherent light from ordinary lamps. Coherent light should make beams much tighter and brighter than those any other source, and that was important for military applications. It would be easy to mark enemy equipment with a bright point of light from a laser beam, giving missiles a clearly identified target to home in on.

Waxing enthusiastic, Gould tossed out his boldest visions as his time ran out. He predicted that a laser beam could be focused down to about the size of a single wavelength of light, only one micrometer—0.001 millimeter—across. Compact disc players do that routinely today, but it was an impressive leap of faith in 1959. Doing that meant power could be concentrated incredibly on a single point. Focus one watt of light onto a one-micrometer spot, and that tiny region would be 10,000 times brighter than the surface of the sun. The power concentrated in a focused laser beam could trigger chemical reactions, or perhaps even give atomic nuclei so much energy they could fuse together, as in a hydrogen bomb. Perhaps eventually, Gould suggested, lasers could be made powerful enough to destroy a nuclear missile before it reached its target.

It was a dramatic finish to a remarkably prophetic sales pitch. Gould's imagination had charged headlong into the future. More than anyone else at the time, he realized the immense potential of a beam of tightly directed light. Townes, Schawlow, Javan, Bennett, and Sanders saw lasers as research instruments and signal sources, generating only the modest powers needed in a laboratory. In that, their visions stayed close to the reality of microwave masers. Gould dared to venture further.

The military scientists were impressed. They knew perfectly well that Gould's bright idea might go nowhere, but that didn't scare them. ARPA had been created to find radical new ways to avoid nuclear doom. Gould held out the prospect of a new beam weapon that could strike at the speed of light. It was a daring idea that might not work, but ARPA's job was to fund daring new ideas in the hope that a few of them would prove to be spectacular successes. Even if the laser could not destroy the most threatening weapons, it might serve other military purposes. Goldmuntz and Gould had come to the right place.

Military agencies had been shopping for energy-beam weapons for decades, with no luck. Most scientists thought death rays were a daft idea, but the brilliant and eccentric Nicola Tesla fueled hopes by dropping hints of incredible inventions as he tried to raise money to build them in the 1920s and 1930s. In 1934, on his 78th birthday, Tesla claimed to have invented a death ray that could instantly kill a million soldiers. Myths grew around Tesla, particularly after some of his papers vanished after his death in 1943. Yet no weapon ever emerged.

The British Air Ministry went on their own quest in the mid-1930s as they uneasily watched the rise of Adolf Hitler. They asked radio researcher Robert Watson-Watt if he could build a radio beam powerful enough to kill or disable enemy pilots before they could drop bombs. He told them no, but said that radio waves could detect enemy planes at night, the basis of radar, which played a major role in winning World War II.

The American military had its own quest at the time. In 1939, a lieutenant colonel told Congress the Army would pay $10,000 to anyone with a death ray that could kill a goat on a 10-foot rope—but that no one had yet claimed the money. Rumors circulated of top-secret trials of mysterious weapons during the war, but nothing emerged. After the war, Theodore von Karman, an eminent professor of aeronautics at Caltech, told military officials it was impossible to build an energy beam to detonate enemy bombs at a distance. Still, military officials kept hoping for a breakthrough.

Gould did not claim he had a death ray. He envisioned a laser beam that delivered watts of power, comparable to the visible light from a 100-watt bulb. Heating a large target would require thousands of times more power. Gould's new idea was to concentrate high intensities of light onto a small and distant spot in a laser beam. Its potential was not lost on military scientists. As curious children, they likely had focused sunlight onto paper with a glass magnifier, charring the paper or setting it afire. Now they could see the potential to apply the same principle to military targets if they could get the laser to work. And if they could get the laser to work, they probably could build bigger ones to fire at more dangerous targets.

Adams was thoroughly impressed, but he couldn't issue a contract without approval from an oversight committee. It took three weeks to get the go-ahead. The panel liked the potential of using laser beams for communications, for marking targets with bright spots for weapons to home on, and for measuring the range to targets for other weapons. Goldmuntz was happy with the panel's approval of a $300,000 contract, but Adams wasn't. He ripped off the final paragraph of the panel's report and wrote in its place, "I, Paul Adams, think we should put $1 million into this program." Goldmuntz was amazed. Military research agencies didn't volunteer more money than contractors requested.

In a matter of days, Adams delivered the final approval, and an acceptance letter full of praise. The total came to $999,008, a lot of money in 1959 and the biggest single deal TRG had ever landed. The company had proposed a series of experiments, to test the laser ideas that seemed most likely to work. One plan was to shine light onto a metal vapor, to see if optical pumping could excite laser action. Another was optical pumping of a solid, looking for signs of laser emission. A third was to pass an electric discharge through a gas, hoping to excite laser action. TRG's proposal called for a small group to perform one experiment at a time, but ARPA was far too excited to wait for the results. The agency told TRG to assign separate groups to each experiment and perform them simultaneously.

Looking back, Gould could understand their excitement. It was the first proposal ARPA had seen for building a laser. "It came out of the clear blue sky that such a thing was actually possible. Of course, ray guns and so on were part of science fiction, but actually somebody proposing to build this thing? And he has theoretical grounds for believing it's going to work? Wow! That set them off, and those colonels, they were just too eager to believe."

ARPA also recognized the crucial importance of looking at a variety of laser media. Gould's proposal, like the Schawlow–Townes paper, outlined the basic optical principles needed to make a laser from a medium that could produce stimulated emission. Gould had listed a variety of materials that might produce stimulated emission in different ways. Each had its advantages, but it was far from obvious which would work and which wouldn't. By studying different types of media separately, ARPA could get the project rolling faster as soon as one of them worked.

The contract accelerated TRG's growth. The company had started in downtown Manhattan, but by the time Gould started work they had grown cramped for space. The final impetus for a move came in 1958 from a fire marshal and TRG's insurance company, who worried about a busy laboratory working with potentially dangerous materials in the densely packed downtown. By the time Gould and Goldmuntz went on the road to sell the idea, TRG had moved to larger quarters on Syosset, Long Island. When they landed the laser contract, TRG had to lease more space in an adjacent building to house new people hired to work on it.

The contract also created stress between Gould and Dick Daly, his immediate boss at TRG. Daly was a take-charge manager, bright and hard driving. He had flown B24 bombers in the Pacific during World War II, and had earned a doctorate in nuclear physics from the Massachusetts Institute of Technology (MIT) before helping develop the first commercial atomic clock at the National Radio Company in the Boston area (fig. 6.1). At 34, he was five years younger than

FIGURE 6.1. Richard Daly (r) with Gerald Zacharias of MIT standing in front of their Atomichron cesium atomic clock, which made the *New York Times* in 1954 when Daly was just 29. (Courtesy of *New York Times*)

Gould, but after only three years at TRG had established a solid track record of project management. He would later found his own laser company.

On the surface, the two men got along well, at least at first. Gould had consulted for Daly while he was still at Columbia. He visited Daly's house and taught Daly to sail. Yet there was an inherent tension between Gould's role as inventor and Daly's role as manager. Gould wanted to run the project he had conceived, but like most scientists he was not a natural manager. Daly was a natural manager, but his take-charge approach could seem overbearing to some scientists. To make matters worse, Daly initially doubted the laser project was feasible. Insiders say Daly went so far as to urge TRG's board of directors to turn down the laser project, but the ARPA money was too good for the company to pass up. Goldmuntz found himself with a delicate balancing act. He needed Daly's management skills. Yet he also needed Gould's technical skills, because no one else at TRG had as clear a vision of the laser.

TRG had submitted its proposal openly, but it was up to the Pentagon to decide whether or not to classify the project. The Schawlow–Townes paper was out in the open, but it said nothing about the potential for high-power lasers. Gould's proposal did, and that made it a candidate for classification. As a military contractor, TRG knew how to deal with security restrictions, which Goldmuntz called "a pain" as he and Gould discussed the possibilities. Gould warned that the problems could be much larger, and for the first time told the TRG president about the ugly political skeletons in his closet from his years on the fringes of American communism.

Goldmuntz was sympathetic and told Gould not to worry. "We'll get you a security clearance. They wouldn't keep you from working on your own project." He had reasons for being optimistic. The anti-Communist hysteria that had cost Gould his teaching job five years earlier was easing. Senator Joseph McCarthy was dead, the Eisenhower Administration had eased some security rules, and the Supreme Court had thrown out others. Moreover, after their negotiations about patent rights, Goldmuntz knew Gould had an all-American capitalistic interest in cashing in on his ideas. Feeling reassured, Gould went off to complete his patent filing. By the end of March 1959, he and patent attorney Robert Keegan had drafted more than 150 pages of text and drawings.

Meanwhile ripples of excitement about the laser proposal spread quickly through the defense research community. Holbrooke sent a copy of the proposal to Bill Culver, a physicist at Rand Corporation in Santa Monica, California. Culver specialized in optics, and he had been among the many people trying to figure out how to adapt Townes's microwave maser ideas to make an optical laser. Rand employed Culver to study potential military uses for new technologies, and he quickly spotted some interesting possibilities for lasers. He flew east on March 30, 1959, carrying his copy of the TRG proposal and a list of eight key questions to ask the scientists he visited.

After first stopping at ARPA headquarters in Washington, Culver visited TRG. He asked Gould what kinds of devices could switch the beam off and on. Gould had already been thinking about that, and suggested putting a switching device between the two mirrors at the ends of the laser cavity. His idea was to let light build up by reflecting back and forth between the mirrors, then switching the interior device to transmit light, so it released the accumulated energy in a powerful pulse. It was a subtle and important invention, and left Culver convinced that Gould had thought through his ideas carefully.

His next stop was Columbia, to visit Townes. As a member of the Air Force Scientific Advisory Board, Townes had seen TRG's proposal to the Air Force, which had left him impressed but puzzled. He didn't think an indifferent student like Gould could have come up with all the ideas by himself. Townes was particularly surprised to see that the ARPA proposal also covered transferring energy between colliding gas atoms. That was Ali Javan's idea, and Javan had not yet discussed it openly. Yet Townes, always discreet and careful to avoid conflicts of interest, said nothing about the coincidence to TRG or Javan.

A phone call to Townes interrupted his discussion with Culver. When Townes hung up, he told Culver the call had come from ARPA, which had decided to classify the TRG proposal as "secret." That posed a problem for Culver, who wasn't supposed to carry "secret" documents without a special letter of authorization. Townes had a secure safe and offered to stash Culver's copy of the proposal inside. Culver declined and took the document home, not wanting to give the TRG proposal to a potential competitor, although he knew Townes had already seen it. It was only the first of the complications created by the security restrictions.

Gould's biggest immediate worry about the classification was the implications for his patent application. Classification puts patent claims into legal limbo. The Patent and Trademark Office can't issue and publish material that's classified, and the inventor can't collect a penny in royalties until security restrictions are lifted and the actual patent is issued. In addition, inventors can file foreign versions of classified patents only in Australia, Britain, and Canada. To avoid that fate,

Keegan worked with Gould and Goldmuntz to split the application into two versions, a short one they hoped could escape classification, and a long one they expected to be classified. They summarized Gould's wide-ranging ideas for exciting atoms in the laser material: identifying promising materials, building reflective cavities for the beam, manipulating the light, and using the laser beam. Keegan filed them on April 6, 1959, the first shot in a legal battle that would continue, with one long hiatus, until the mid-1980s.

Meanwhile, TRG applied for a clearance for Gould. The final ARPA contract arrived in mid-May, spelling out security requirements. Status reports, engineering notes, computations, and some other paperwork were ruled "confidential," the lowest level of classification. However, experiments, test results, technical reports, and final conclusions were classed at the higher "secret" level.

Security clearances are complex matters, with a military bureaucracy all to themselves. ARPA had some influence in the process, but the agency did not have the final decision. Paul Adams at ARPA promised to help get a clearance for Gould, but the decision was out of his hands.

Gould's lack of a clearance quickly started getting in his way on the laser project. He and Goldmuntz flew to California to talk with Culver and other Rand scientists about the project. After they settled into a Rand conference room, Culver's boss walked in and snatched Culver's copy of the TRG proposal without explanation. Puzzled, Culver followed him out of the room—and was confronted by Rand's security chief, who scolded him because Gould had not submitted his nonexistent security clearance.

"But it's his work," Culver protested.

That didn't matter to the security chief. Rules were rules, and rules had to be obeyed. "It's classified, and you can't discuss it with him."

Complications multiplied as TRG got the laser project going. The company shifted some people from other projects, but it also hired new people to work on the laser. Gould did much of the interviewing, and new hires got the impression that Gould was in charge. Yet without a clearance, Gould couldn't go inside the secure area TRG had set up for classified research in its second building. Goldmuntz put Daly in charge of the classified operation.

TRG started pulling strings to push Gould's clearance through the system. For legal help, Goldmuntz turned to a childhood friend with a one-man Washington law practice, Adam Yarmolinsky, who drew in fellow lawyer Harold Leventhal. Both were rising stars in the capital. Yarmolinsky later played important roles in the Kennedy, Johnson, and Carter administrations, and Leventhal would become a Federal appeals court judge.

They quickly realized Gould would be a hard sell. Although the worst excesses of McCarthyism were gone, the security establishment was slow to change. Straight-arrow military bureaucrats had no sympathy for Gould's leftist politics and Bohemian life-style. Gould was not only divorced from his first wife, but he had openly lived with both her and his second wife before they married. That was not done in 1950s America. Personal references were another problem. Security clearances depended heavily on them, but Gould hadn't gotten along well with the conservative side of the physics establishment. Yarmolinsky worried that two of Gould's best personal references were men who wore beards, which at the time were considered suspiciously subversive by the clean-shaven establishment. Gould's leftist record went back to 1945, and while he had sworn off communism, he wasn't about to repudiate his liberal beliefs. A college radical might be repackaged with a shave, a haircut, and a clean suit. Gould posed a much bigger challenge.

7
A SPREADING INTEREST
IN THE LASER IDEA

IT TOOK TIME FOR THE APPEAL of the laser to spread. In early
1959, the laser was the kind of far-out research project that belonged at ARPA,
Bell Labs, or a university. It looked good on paper, but lots of things that look
good on paper never work. The laser was still in the basic research stage. The real
action and the serious money were in the development of microwave masers.

Research and development are often lumped together, but they differ in
important ways. Research is the exploration of new ideas, and its results are
never certain. Development is building things using ideas that have already been
tested, and the results can usually be predicted, although they may cost more
than expected. The optical laser was research. The microwave maser had become
development, which usually takes more people and more money than research.

Microwave masers had become a hot topic because they promised a way to
amplify microwaves with very low noise levels, so the signals were sharp and
clear. Townes was intrigued with their potential for radio astronomy, which
received only the faintest whispers of signals from the distant cosmos. Bell Labs
wanted microwave masers to build communication systems that could carry signals further. The military wanted microwave masers to make their radars and
communication systems more sensitive.

Outside of Bell Labs, the military paid for most of the American research and development on microwave masers. The basic research, often done at universities, generally remained unclassified. Most development was done at a network of defense contractors, which lured bright young scientists with attractive salaries. Virtually all of the scientists were male, and they had to put up with security restrictions that could be intrusive, some were pushed into dropping girl friends the FBI thought were "communist sympathizers." But America was still in the conformist 1950s, the money was good, and the work was interesting. Scientists like Gould, who dabbled in leftist politics, often avoided military research and security clearances. Many others saw military research as a golden opportunity to be paid well for working in well-equipped laboratories on projects of national importance.

Intellectual challenge was another attraction of military development programs. Working on projects like microwave masers posed the type of challenges that bright scientists enjoyed. Their projects introduced them to new concepts and materials, stimulating their native curiosity and giving them new ideas for future projects. The project that opened the door for Theodore Maiman was a contract landed by Hughes Research Laboratories in California to develop a better version of the ruby solid-state microwave maser.

Hughes Aircraft was the creation of Howard Hughes, the best known and most eccentric industrialist of his generation. A master wheeler-and-dealer, Hughes made his fortune producing oil-well equipment, and earned his fame producing movies. Fascinated by aviation, he formed Hughes Aircraft in the 1930s. The company built the world's largest airplane, the wooden "Spruce Goose," which Hughes himself flew on its only flight just after World War II. Hughes Aircraft prospered as a Cold War defense contractor by solving tough problems in avionics and electronics, growing from 1000 people in 1948 to 20,000 in 1955. By then, Howard Hughes had largely withdrawn from the scene. Lawrence A. "Pat" Hyland, who became general manager of Hughes Aircraft in 1954, never met the industrialist in person, but did receive calls and notes from him. Long-time employees believed Hughes visited company offices late at night, when no one was there to see him.

Given essentially complete control, Hyland built a top-level research laboratory that initially focused on microwave and electronic equipment. Among the people hired during the build-up were Harold Lyons, who in turn hired Maiman.

Lyons made his scientific reputation by developing the ammonia-beam atomic clock at the National Bureau of Standards in 1948 (see fig. 7.1). Atomic clocks keep time by making atoms or molecules oscillate at extremely precise frequen-

FIGURE 7.1. Harold Lyons (right) looks a bit glum as he poses in 1948 with the ammonia beam atomic clock and Edward U. Condon, director of the National Bureau of Standards. (National Bureau of Standards photo, courtesy of AIP Emilio Segre Visual Archives, *Physics Today* Collection)

cies set by their internal energy levels. Although that sounds like esoteric academic research, precision timing is also a matter of military concern, because it's essential for navigation and guidance of ships, aircraft, and missiles. The physics are closely allied to microwave masers, and Townes had interested him in using ammonia beam masers as frequency standards.

At NBS Lyons headed a section developing microwave frequency standards, but his ambitions went beyond the federal bureaucracy. At 42, he headed west in 1955 to manage research on atomic and molecular resonances—a field that included microwave masers and atomic clocks—at the fast-growing Hughes, where he saw a better chance to climb the management ladder. He soon negotiated a contract to build gas masers to generate precise microwave frequencies for the Army Signal Corps. That made Hughes one of the first aerospace companies to be involved in microwave maser development.

Ted Maiman was among the cadre of young scientists hired as Lyons expanded his group. His family's roots were Jewish; the name Maiman can be traced back to the twelfth-century Jewish philosopher Moses Maimonides. The bright and restless son of an electronic engineer who spent his entire career working for AT&T, Maiman was born in 1927 in Los Angeles. He grew up in Denver, learning electronics under his father's tutelage. The younger Maiman landed a job as a junior electronic engineer at 17 after finishing high school in three years, then near the end of World War II signed up for the Navy, where he entered a radar program. When he returned from the service, he studied electronic engineering at the University of Colorado. Stanford's highly selective physics department initially rejected him for graduate school, but he got into Columbia, which he didn't like. He tried again for Stanford's physics department, and after they rejected him he hitchhiked from Denver to Stanford and talked his way into the electrical engineering program. After picking up a master's in engineering, he tried again to get into Stanford's physics department. The third time was the charm, and the determined young Maiman enrolled in the doctoral program.

His thesis advisor was Willis Lamb Jr., who after 13 years on the Columbia faculty had moved to Stanford in 1951. At Columbia he had studied the spectrum of hydrogen, work that earned him the 1955 Nobel Prize in physics, which he shared with Polykarp Kusch. Lamb needed to build a new lab at Stanford, and Maiman's electronic engineering skills were a godsend. Although physics may sound close to electronics, the fundamental approaches differ. Physicists try to demonstrate a principle; electronic engineers make things work. Maiman was very good at designing and building his own equipment. It was not a common skill among physics students, but it helped Maiman greatly in making the difficult measurements on helium atoms that Lamb suggested as his dissertation topic. After struggling with the original set-up, Maiman devised his own apparatus for the measurements, and following some fine-tuning, he had a nice, clear signal.

Lamb wanted him to stay after that success, but by 1955 Maiman was getting restless after four years in the basement of the Stanford physics department, and he used his hard-earned savings to book a cruise around the world. He had to finish writing his thesis first, and Lamb also insisted that before he leave he spend some time helping another graduate student, Irwin Wieder, who was measuring some different properties of helium using some of Maiman's equipment. But Maiman met his deadlines and sailed on schedule. He enjoyed the adventure, but exhausted $3500 of savings and had to borrow $500 from his father to get home. Wanting to get out of academia, Maiman landed a job in the fall of 1955 at Lockheed in the Los Angeles area, then jumped to Hughes Research Labs in January 1956.

Lyons had assembled a bright and colorful crew of people, and Maiman plunged happily into military sponsored research. However, after a couple of years Lyons put Maiman in charge of a different type of project, a contract from the Army Signal Corps to build a more practical version of the ruby microwave maser that Chihiro Kikuchi had demonstrated at the end of 1957. The Army was looking for a low-noise microwave amplifier to put in an airplane. Ruby looked like a good choice because it was durable and easy to obtain, but Kikuchi's version wouldn't fit in a plane.

Ruby had to be cooled far below room temperature to keep the chromium atoms in the proper states for microwave maser operation. Kikuchi put the crystal in a flask of liquid helium, four degrees above absolute zero. Left by itself, the liquid helium evaporated very quickly, so the flask of helium had to be placed inside another flask containing liquid nitrogen. Then the whole assembly had to be mounted inside a strong magnet to produce the right energy levels in the ruby. When Kikuchi put the thing together, it weighed 2.5 tons and was as big as a desk—but it worked.

What Kikuchi had done was a physics experiment. The Army wanted a practical device. There's a big difference between the two. A physics experiment is supposed to illustrate a concept, which may often be fairly simple. However, the apparatus used to demonstrate that concept doesn't have to be simple, and usually it isn't. Like the ion-beam experiments at Columbia, physics demonstrations tend to be cumbersome, and require the careful attention of graduate students to work properly. Practical devices are supposed to look simple, and work easily, but the underlying concepts may be far more complex. The gap can be as huge as the one between the first steam-powered three-wheeled car built in 1769 by

Nicholas-Joseph Cugnot in France and the Mark-7 Jaguar that Maiman tended for Lamb while the professor was visiting at Harvard.

Initially, Maiman was disappointed. He had wanted to do research, not build a device, which he considered an engineering project. He had begun thinking about how to make an optical laser, which he considered a far more interesting prospect. But he was newly married, with his first child on the way, and he didn't feel he had much choice about the matter. After talking with his immediate boss, George Birnbaum, Maiman agreed to go ahead.

Maiman started out knowing little about ruby, but he caught on fast and came up with a far more practical version of the ruby microwave maser. His key innovation was to put a 12-ounce magnet inside the liquid helium flask instead of putting the flask inside a massive magnet. Inside the magnet he placed a ruby slab seven millimeters square and three millimeters thick, which he painted with a specially formulated silver paint that reflected microwaves. The finished package weighed only 25 pounds, and worked much better than expected as the front end of a radar receiver, much to the delight of the Signal Corps. The experience taught Maiman how careful engineering could make a dramatic difference in performance—and that cryogenic cooling could be a nuisance.

He left most of the actual assembly to his young assistant, Irnee D'Haenens, the amiable son of an Indiana service-station owner. D'Haenens had started working for Maiman while he was working on a master's degree in physics at the University of Southern California, and took a full-time job when he finished. He lacked Maiman's intensity, and that may have helped the two get along well. Hughes offered a pleasant environment for bright young scientists, and the generous benefits were a lure for D'Haenens. Employees who stayed with the company several years could qualify for employee fellowships to go back to school, and with three young children that was the only way he could afford to return to school to earn a doctorate.

The next step was redesigning the ruby microwave maser so it could operate at the higher temperature of liquid nitrogen, 77 degrees above absolute zero. That eliminated the liquid helium, and cut the weight to a mere four pounds. Yet in the end even that wouldn't overcome the drawbacks of having to cool the ruby crystal. The Army decided to go with other sensitive microwave amplifiers that didn't have to be cooled to cryogenic temperatures.

The microwave ruby maser experience left Maiman with useful insights that would prove important when he turned his attention to lasers. Microwave masers were attractive as amplifiers because they offered very low internal noise levels. That noise increased with temperature, so moving from cumbersome liquid-helium cooling to simpler liquid nitrogen raised the noise levels. Searching for ways to reduce that noise, Maiman examined the effect of how the chromium atoms were excited to produce a population inversion. He found that increasing the frequency of the microwave signal that excited the chromium atoms should cut down on the noise.

The immediate problem with that idea was that he couldn't just turn up a knob. His microwave source was already near the high end of the available frequency band. However, Maiman had also heard of the hot new technique of optical pumping that had already intrigued Gould and Townes. He knew light waves have frequencies tens of thousands of times higher than microwaves, so exciting ruby with light should reduce the noise level below what he saw with microwaves. Maiman and D'Haenens tested the idea by focusing light onto ruby samples. Bright pulses of visible light excited many chromium atoms to jump from the ground state to an excited state where they stayed for a few milliseconds before dropping back down to the ground state. Those results looked encouraging for microwave masers, and encouraged Maiman to think of going a step further. If light could excite ruby enough to emit microwaves, could it also excite the chromium atoms to emit visible light?

Curiously, similar thoughts about the optical properties of ruby were going through the mind of Irwin Wieder, who had followed in Maiman's footsteps at Stanford. Like Maiman, Wieder had gone to industry after finishing his doctorate under Lamb, but Wieder had headed east, to work at Westinghouse Research Laboratories in Pittsburgh. Westinghouse had an Air Force contract to study solid-state microwave masers, and one of the materials Wieder studied happened to be ruby.

A native of Cleveland, Wieder was a couple years older than Maiman, but wound up behind him in school after losing time to World War II and college policies that discouraged admission of orthodox Jews. When he came to Stanford, Willis Lamb started him out helping Maiman, a traditional way of getting new students involved in laboratory work. It was useful experience for Wieder. Maiman and Lamb were both sons of telephone engineers who had learned practical skills from their fathers, but Wieder's father—like Gould's—was not at all handy with tools. After learning the ropes in the laboratory, Wieder started building his own apparatus, although he was able to use some of Maiman's for his thesis project, which involved different measurements of helium. He graduated in 1956, as the restless Lamb headed for a prestigious faculty post at Oxford University. In August, Wieder started at Westinghouse, then one of America's largest industrial corporations, with decades of experience in building electrical equipment.

After Wieder arrived, Westinghouse's little maser group chatted over lunch about new ideas, including the possibility of making masers that emitted light instead of microwaves. They finally concluded it wouldn't work because they couldn't see a way to selectively excite the states needed to make a population inversion. If they tried to excite it the same way as a three-level microwave maser, an optical laser would require thousands of times more energy. That didn't look practical.

Wieder's real job was to develop better microwave masers, but he was not ready to give up on the idea of using light. Looking for new ways to excite atoms in a solid, he hit on a new variation on optical pumping that he first outlined on January 15, 1958, and later wrote up for the Westinghouse patent department. The idea was to concentrate or "funnel" light energy emitted over a broad range of wavelengths into a much narrower band, then use that narrow range of wavelengths for optical pumping. His first goal was a microwave maser, but in principle the idea would work for an optical version as well.

The idea of optical pumping that Rabi had introduced to Gould was based on exciting atoms with light at a narrow range of wavelengths. Its big advantage was that the narrow range of wavelengths selectively excited certain gas atoms to specific energy levels. (No one had yet demonstrated optical pumping in solids.) That made it a great research tool. However, it was woefully inefficient. The special lamps that emitted a single wavelength didn't produce much light. The atoms in a gas are far apart, and they didn't absorb much of that small amount of light.

Wieder wanted a more efficient way to produce a narrow band of wavelengths, and he thought that fluorescence from solids might provide the answer. Fluorescence occurs when atoms absorb light photons and almost immediately emit light at a longer wavelength. It's spontaneous emission, not the stimulated emission of a maser or laser. Some solids absorb light energy over a broad range of wavelengths, and emit fluorescence over a much narrower range. One of them is ruby, in which chromium atoms absorb ultraviolet, violet, blue, and green light and emit deep red fluorescence. Wieder wanted to tame this phenomenon to serve as a "funnel" that would concentrate energy from the wide range of wavelengths that could excite chromium atoms into the narrow range of the red fluorescence. He hoped to use that red fluorescence to optically pump a microwave maser.

That two-stage approach might seem cumbersome, but Wieder saw it as a way to simplify optical pumping. Ordinary light bulbs emit light across a broad range of wavelengths, and his scheme was really a way to capture all that light energy and funnel it into the narrow range that gave optical pumping its selectivity. That would let him produce a population inversion by exciting a material with white light. The emission line would have to match an absorption line of the maser material, but Wieder figured he could do that by using the same material in the fluorescence source and the maser, and ruby looked like a good candidate. It was an innovative idea and he's still proud of it.

When Westinghouse negotiated a microwave maser research contract with the Air Force, Wieder included his proposal for a solid-state pump source. The contract specified the goal as developing microwave masers, but it was broad enough to give Wieder time and money to study the optical properties of ruby. Experimental progress was slow. Wieder and his colleague Bruce McAvoy started by illuminating ruby samples at room temperature. When that didn't have much effect, they tried cooling the crystals in liquid nitrogen.

McAvoy suggested another way to increase the light output—exciting the ruby samples with pulsed light sources, which can briefly attain peak intensities far higher any steady light source. Pulsed masers had received little attention, but the idea seemed intriguing. McAvoy decided to start by illuminating the ruby with a high-intensity photographic flashlamp. The budget didn't include money for that experiment, but he found enough money to buy a standard commercial flash. The special power supply used with the lamp was more expensive, but Westinghouse had plenty of electrical equipment lying around that he could scrounge. He borrowed a bank of capacitors, devices that can store electricity over a period of time, then discharge it very quickly. To control them, he found a big old mechanical switch. The combination looked like it should deliver enough power to excite the ruby.

The capacitor bank did indeed deliver plenty of energy. Unfortunately, it was considerably more than the flashlamp was rated to handle, and the lamp exploded.

Nobody was hurt, but the event attracted the unwelcome attention of the laboratory safety officers. The explosion itself was bad enough. But when they inspected the jury-rigged switch and capacitor bank, they found that a person using them could easily contact voltages packing enough of a wallop to kill.

Building an enclosure might have solved the problem, but the capacitor bank was already too bulky to fit into Wieder's small lab, so he gave up on pulsed light sources. Instead he turned to the brightest tungsten lamps he could find, which generated a steady light, and set out to measure how efficiently it could excite fluorescence from ruby.

The flashlamp fiasco didn't go into the quarterly progress reports the government required, but other details did, including the idea that optical pumping might be used to extend the maser principle to infrared wavelengths. The details were unclassified, and Wieder sent copies to a list that included Charles Townes, Ted Maiman, and a large number of military researchers.

Others also were thinking about making the jump from microwave masers to optical versions. It seemed a logical extension of developments such as the three-level microwave maser. A few took steps toward the laser concept, but didn't see the whole thing. In the U.S., Princeton University physicist Robert H. Dicke patented a "molecular amplifier" with a pair of mirrors that reflected light between them. In the Soviet Union, Prokhorov published a short letter that proposed a "molecular amplifier" that also used a pair of mirrors. Yet both fell far short of the Schawlow-Townes and Gould proposals, and most other ideas were largely lunchtime speculation that never led to any careful analysis or serious experiments.

American scientists started taking the laser idea more seriously after they saw the analysis by Schawlow and Townes. Nicolaas Bloembergen was interested, but he decided not to try building a laser. Harvard didn't have the depth of support facilities and people that he felt he needed to compete, and Bloembergen was not about to enter a race he didn't think he had a good shot at winning. IBM did have resources, and Bill Smith, a research manager at the company's stylish new research laboratory in Yorktown Heights, New York, decided the laser would be a good research topic for two young physicists working for him, Peter Sorokin and Mirek Stevenson. They were logical choices. Sorokin, the son of an eminent sociologist, had earned his Ph.D. at Harvard under Bloembergen. Stevenson had earned his under Townes at Columbia. But like everyone else, it took them a while to get going because no one knew exactly what it would take to make a laser.

The Americans had little idea what was going on overseas, particularly in what seemed the most serious competitor, the Soviet Union. The Iron Curtain, language, and geography created high barriers to communications between the U.S. and the Soviet Union. Basov and Prokhorov's microwave maser work was public, showing the Russians were up to something, but exactly what they were doing was an open question. Americans had to wait months for translation of the leading Russian journals, and the English versions often left many questions unanswered. Informal reports, conference papers, patents, and more obscure journals were never translated, and much research was never published openly at all. Much of Soviet science was a mystery, even to the intelligence agencies paid to decipher it.

In reality, the Russians had good ideas, but a shortage of sophisticated instruments and equipment made it hard for them to translate those ideas into successful experiments. After the microwave maser was demonstrated, Fabrikant and Butayeva tried to excite population inversions by firing pulses of light or electricity into various gases and vapors. In 1959, they claimed the first success in amplifying light. They optically-pumped a mixture of mercury vapor and hydrogen gas, and found that the strength of blue and green lines of mercury at 435.8 and 546.1 nanometers increased by about ten percent. The idea was a reasonable one. They thought the light would excite the mercury atoms, and the hydrogen molecules would remove mercury atoms from a lower energy level, creating a population inversion. However, it turned out they were fooled by their instruments. The sensitive photomultiplier tube they used to measure the light had responded too strongly, giving the wrong reading. The incorrect results would have encouraged American researchers—and unsettled the Pentagon— but word of them didn't trickle out of Russia until after the first laser had been demonstrated.

8

A PAUSE TO COMPARE NOTES

EXCEPT FOR BELL LABS, all laser research through early 1959 was widely dispersed, performed by small groups or individuals in much larger organizations. The TRG contract was changing that, but in the spring the company's newly funded laser project was still getting on its feet, shifting people from other projects, hiring new people, setting up experiments, and waiting for Gould's clearance to come through. The little group at Columbia had some money, but others had no direct support for working on lasers, and some, like Maiman and Wieder, were being paid to work on microwave masers.

As spring arrived, researchers' thoughts turned to staking their claims on new ideas and picking up useful information from others. Bell Labs had a more subtle agenda. Management worried that the Pentagon would try to extend security restrictions beyond the TRG contract to include other laser research. To forestall that, Bell management pressed Javan and Sanders to get their ideas into the open by publishing papers in scientific journals and giving talks at research conferences.

Their first public talks were at a small and rather informal conference held June 15–18 at the University of Michigan in Ann Arbor. The main subject was optical pumping, which had become a hot topic because of its applications in basic research. But some laser-related research was slipped into a "miscellaneous" session chaired by Art Schawlow.

John Sanders had only modest progress to report more than half-way through his nine-month sabbatical at Bell. His strategy was to get his experiment up and running, hoping that passing an electric discharge through a tube containing helium gas and a pair of mirrors 7.5 centimeters apart would yield a population inversion on a helium transition at a red wavelength of 0.668 micrometer. The excited helium atoms normally stayed 30 times longer in the upper energy level than in the lower level, so he hoped they would accumulate in the upper state while the lower one emptied out. If that worked, spontaneous emission by a few excited helium atoms might trigger a cascade of stimulated emission from the inverted population of helium atoms, which would grow as the light bounced back and forth between the two mirrors. But by the Ann Arbor meeting, Sanders could see that it wasn't going to be that easy. A competing process called radiation trapping was putting more atoms into the lower energy level, preventing a population inversion. Sanders said he planned to try more experiments under other conditions to see if he could do any better. A couple of weeks before the meeting he submitted the same results to *Physical Review Letters*, which published his short letter in mid-July, putting his idea firmly into the open.

Javan also described his research. He and Bennett were in the midst of a much more meticulous analysis of what happened to gas atoms in an electric discharge. Their experiments and calculations had taken them a critical step beyond Sanders, to understanding how to keep radiation trapping from undermining efforts to produce a population inversion, which Javan still called a "negative temperature." The central problem was a conflict between the requirements for the atoms to collect energy from the electric discharge and to maintain a population inversion. Atoms collect energy when they collide with electrons in the discharge, so the gas density should be high in order to collect energy efficiently. Yet increasing the gas density strengthens the amount of radiation trapping, which destroys the population inversion.

The solution to the problem was to transfer energy differently within the gas. Sanders was using what physicists call "collisions of the first kind," when electrons hit atoms. Javan and Bennett had instead turned to "collisions of the second kind," in which different atoms hit each other. Both types of collisions transfer energy, but collisions of the second kind let Javan and Bennett use a mixture of a lot of one gas and a little of another. The more abundant gas collects most of the energy from the electrons. Unexcited atoms of the less common species can then pick up energy if they collide with excited atoms of the more common species. Getting the process right requires finding a pair of gases with energy levels that match, and mixing a lot of helium with a little neon was looking better and better. Javan also wrote up his ideas for *Physical Review Letters*, going into considerably more detail than Sanders, another step toward keeping laser research unclassified. Bennett was annoyed that Javan neglected to list him as a coauthor, but he couldn't do much after the fact.

After his patents had been safely filed, Gould was free to talk at Ann Arbor, within the constraints imposed by the ARPA contract and his efforts to secure a clearance. Like Javan, Gould was staking out intellectual turf. He staked his claim to the word "laser" by titling his talk, "The LASER: light amplification by stimulated emission of radiation." He traced the origins of his work to the fall of 1957, when he started his first notebook, rather than to the formal proposal TRG submitted at the end of 1958 or the patent application filed in April 1959. He said TRG's classified research program was studying seven different ways to deposit energy in a laser material, and four different optical designs, but he didn't give details.

The meat of Gould's talk was on a safely unclassified topic, how amplification and oscillation should affect the range of wavelengths emitted. Like Schawlow and Townes, Gould showed that the process would produce a wavelength peak much sharper and narrower than spontaneous emission. That's a natural consequence of the nature of the process. The gain, or degree of amplification, varies with the wavelength. Each time light passes through the laser medium, it is amplified most at the wavelengths where the gain is strongest. That builds the stimulated emission peak many times higher and narrower than spontaneous emission. It's a key feature of laser light, and measuring that wavelength peak is one way to check if a device is operating as a laser.

Happy with the hefty ARPA contract, Gould drew chuckles by boasting of funding from "Uncle Cornucopia." But Schawlow brought down the house afterward. The laser was likely to be used mainly as an oscillator generating light rather than as an amplifier, Schawlow said, and with mock solemnity suggested a change in acronym. An optical oscillator was really "Light oscillation by stimulated emission of radiation," Schawlow duly noted, making it a "LOSER" rather than a "LASER."

Schawlow's inner imp had taken charge, and he was laughing by the time he reached the punch line. He was a jovial soul, a trait that made him many friends over the years. Yet while Schawlow enjoyed his fun, he took his science seriously. Beneath the surface, a split was growing within the small, young community of laser researchers, and Schawlow was caught up in the underlying tension. Two sides claimed the idea was theirs, and the litmus test to separate them was their choice of names.

Schawlow and Charles Townes called their concept the "optical maser." They clearly hoped the optical maser would become an important invention, but the label marked it as a variation on the (microwave) maser. That linked it to Townes's earlier development of the maser. Typical of Townes, they had played by the rules of the scientific establishment, carefully documenting their ideas and publishing them in a formal journal. Bell Labs and Columbia were firmly in the establishment circle with them, so were many other scientists who had studied at Columbia or worked at Bell Labs.

Gordon Gould chose his own name for his idea, just as he tried to steer his own course through life. The word "laser" said the new concept was kin to the maser, but also that it was an invention in its own right. His new term had the advantage of being fresh, simple, and compelling. Its two syllables rolled off the tongue more smoothly than the five of "optical maser." It sounded bold and brash, fitting with the upstart TRG, and with ARPA's charter to invest in risky new ideas with the potential for big payoffs. They were the outsiders, challenging the establishment, and gradually building up their own circle.

Townes and Schawlow probably had expected Gould to fade away, especially after he dropped out of Columbia. Gould's erratic work habits had never impressed the methodical Townes. "He seemed bright enough, but not terribly interested," Townes said later. Yet somehow Gould had kept going, and with TRG, had landed a fat military contract that put the upstarts in a position to challenge two of the establishment's premier physics laboratories, Bell and Columbia. Moreover, the military interest was threatening to envelope laser research in a veil of security restrictions. The striking similarity of the proposals for the optical maser and the laser encouraged suspicions in both directions. It was natural for Townes to suspect a student he considered not particularly diligent might have filled his notebooks with borrowed ideas. Schawlow also seemed to suspect Gould, perhaps because he was close to Townes. Likewise, it was natural for Gould, scarred by earlier encounters with the establishment, to suspect Townes of borrowing his ideas. Each of the three men knew how long and hard he had worked to develop his own ideas, so it wasn't easy to accept that someone else had developed very similar ideas completely on their own.

Other news presented at the Ann Arbor meeting would also play a role in shaping the course of the laser race. Irwin Wieder came to report a modest success in his experiments on the optical-pumping of ruby. He had illuminated a ruby crystal with a tungsten projector lamp. The crystal had absorbed about ten watts of green and blue light, and Wieder had detected the red fluorescence he was hoping to find. He didn't try to measure the amount of red light very carefully, because it wasn't crucial to his goal of using the red light to pump a ruby microwave maser, but he estimated the power at about one watt. If that number was right, it meant that about one out of every ten photons from his lamp triggered chromium atoms in the ruby to emit red light. Physicists call that number the quantum efficiency, because it tells how efficiently photons (quanta of electromagnetic energy) can transfer their energy.

That power and efficiency had impressed Wieder, and he mentioned the power he had produced when he called Peter Franken, the young University of Michigan professor who chaired the conference. Wieder succeeded in talking his way onto the program, but Franken didn't believe that Wieder could be getting as much as one watt of output. Franken had been studying optical pumping, but he was working in gases, where the power levels were milliwatts to tens of milliwatts, not in solids where the atoms were packed more tightly together and could emit more light.

Franken hadn't insisted on a change, but Wieder was young and insecure. He wanted to be on the program, and with deadlines tight he didn't have time to make a new set of detailed measurements. Wieder cut the estimate of ruby output he gave at the meeting to 0.1 watt. That reduced the estimated efficiency to one percent, although that number appeared only in a paper that was in press at the time of the meeting. It was a deliberately conservative estimate, because he couldn't document the case for higher efficiency. If it had been important for his work, Wieder could have gone back and made careful measurements, but he didn't see it as necessary for him to make an optically-pumped microwave ruby maser. He had clearly produced more than the 0.01 watt he could get at that red wavelength from a lamp, and that was enough to show that the red ruby light was promising.

The fluorescence seemed bright enough that optical pumping should have excited the microwave maser, but it wasn't working. Measuring quantum efficiency wouldn't help, because quantum efficiency is inherent in the material; Wieder knew he couldn't change it. He concentrated on factors he could adjust, changing light sources and the optics that focused light onto the ruby. He tried cooling the fluorescence-emitting ruby crystal to liquid nitrogen temperature and the microwave maser crystal to liquid helium temperature. He refined his experiments, so the signal should be 100 times higher than the background noise, but still saw nothing. At the time of the Ann Arbor meeting, he was still trying to understand what was going wrong.

The quantum efficiency of ruby fluorescence was very important to Art Schawlow, who was still considering ruby as a laser material. Fluorescence efficiency is closely related to the probability of stimulated emission. Both depend on how likely the input light is to excite chromium atoms to the desired higher-energy state. The standard isn't too demanding for fluorescence alone; all that is required is exciting some chromium atoms so they can release red light afterward. But it's critical when trying to make a laser from pink ruby. A laser can work only when more chromium atoms are in the upper laser level than in the lower one. In pink ruby, the lower laser level is the ground state, and the only way to have more chromium atoms in the upper level is to excite more than half of them. If the efficiency was limited to one percent—or even ten percent—there was no hope of inverting the population. Wieder didn't put a number on efficiency in Ann Arbor, but Schawlow could figure it out if he divided the 0.1 watt of fluorescence by the amount of white light absorbed.

After Wieder talked, Schawlow pointed out that red ruby has other emission lines arising from interactions between nearby chromium atoms that don't exist in pink ruby because the chromium atoms are too few and far between. Yet the most important message Schawlow took home was Wieder's estimate of low efficiency in exciting the primary ruby fluorescence. Others at Bell had said as much earlier, but Wieder's estimates were important as confirmation. To Schawlow, they were one more nail in the coffin of pink ruby as a practical laser material. Had Wieder been the first to report low efficiency, Schawlow might have tried his own measurements, but he didn't think he needed to because others had said the same thing.

The Ann Arbor meeting was far from the last word on the state of laser research. Charles Townes hadn't attended because he was busy organizing another meeting for the Office of Naval Research, to cover microwave masers, atomic clocks, and related concepts including the optical maser. As a key mover and shaker behind maser development, Townes was a logical choice. He assembled a top-level steering committee that included Bloembergen of Harvard, George Birnbaum of Hughes, Bob Dicke of Princeton, and Rudi Kompfner of Bell Labs. The Columbia physics department staff, his graduate students, and a secretary hired with a Navy grant handled the logistic details.

Townes also had other things on his mind. After years of serving on military advisory panels, he had been asked to serve as vice president and research director of the Institute for Defense Analyses, a Washington think tank that a group of universities ran for the Pentagon. It was a challenge that interested Townes. Science had become increasingly important in the aftermath of Sputnik and in the midst of the nuclear arms race. The government needed top-level scientific advisors who could understand the fast-changing world. Most scientists were reluctant to go to Washington because they considered administration a chore and had little love for politics. Another top physicist, John Wheeler of Princeton, had already turned down the job. The wounds left by McCarthy era witch-hunts were still raw. Rabi thought Townes had to believe the country was in desperate shape if he was willing to go to Washington. Townes felt it was his duty to take on the job.

It was also a step up the administrative ladder, and Townes nursed ambitions beyond the laboratory. He suspected that the maser and laser might be worth a Nobel Prize, but he thought the essential steps toward the laser had already been taken. Nobels often are awarded to the originators of important new ideas, even if they don't see them through to completion in the laboratory. "I felt it did not really matter who actually built the first one. The ideas were there," he wrote in his autobiography. Townes discussed the opportunity with his family and a few colleagues. As he approached his 44th birthday, Townes decided to accept the IDA post, which he felt could have an impact beyond the little world of physics.

His students had no inkling of his plans until early summer, when Townes posted a notice asking them all to come to a meeting. Isaac Abella immediately knew something unusual was happening; Townes normally met with individuals or a few people working on related projects, not with all of his students. At the meeting, Townes dropped the bombshell that he would be taking a two-year leave to work at IDA.

Abella initially felt devastated. Doctoral students work closely with their dissertation supervisors. One of the student's worst nightmares is that the professor might take off, leaving them stranded with years invested in a project they can't complete. Townes promised to keep in touch with students who were nearly finished their dissertations, but suggested that others might want to find different advisors. Privately, he assured Abella and Herman Cummins that he would continue working with them on the optical maser, which clearly had a high personal priority for him. That also showed in his choice of a temporary replacement for the fall semester, a British specialist in optics, Oliver S. Heavens. He hoped that Heavens's optical expertise could provide a vital push for the project.

Townes forged ahead with planning the maser and atomic clock meeting for the Navy. Called "Quantum Electronics–Resonance Phenomena," the meeting was held September 14–16, 1959, shortly after he moved to Washington. The site was Shawanga Lodge, a resort hotel in the Catskills, about 90 miles northwest of Manhattan, where the rates dropped after Labor Day. The organizers invited almost everyone who was anyone in maser and laser research, although they missed Bill Bennett, whom Javan had neglected to list as a coauthor in his papers, and only Gould and Maurice Newstein came from TRG. Over 160 scientists showed up, not counting more than a dozen students and others listed as "conference staff." Basov and Prokhorov came from the Soviet Union, on their first visit to America. Other scientists came from Britain, France, Germany, Holland, Switzerland, Japan, and Israel. Rudi Kompfner and leading communications engineer John R. Pierce headed a formidable delegation from Bell Labs. Eight attendees eventually received Nobel Prizes, and many others became leaders in laser research.

Today, the Shawanga Lodge meeting is fabled as marking the birth of laser research. Ironically, virtually all of the 66 papers presented dealt with microwave masers or other research rather than lasers. Only two, by Schawlow and Javan, specifically covered optical masers. Yet laser ideas were in the air.

The sense of keen competition in the growing laser race made some scientists wary of revealing too much. Bell Labs, worried that the government might impose security restrictions on its research, was an exception, although John Sanders, nearing the end of his stay, didn't talk about his work. Gould didn't give a paper, perhaps because of security restrictions, and limited his role to raising questions after others talked. Other groups had nothing to report, or were still formulating their projects.

Javan had little to add to his Ann Arbor talk. After he finished, Gould noted that spectroscopists nearly 30 years earlier claimed to have inverted the population of sodium vapor by transferring energy from mercury atoms. Townes's students dutifully recorded Gould's addendum for the published conference proceedings, but no one recorded private conversations between researchers sharing ideas in hallways and over meals.

Art Schawlow showed up late. His talk was scheduled for the final day, and he had fallen behind on the written version he had promised, so he spent the first two days of the meeting at home in New Jersey, writing furiously. In his talk, Schawlow touched the important bases of optical maser research. For the first time, he clearly predicted that bouncing stimulated emission between a pair of mirrors would produce a beam that could concentrate energy onto a tiny spot. He envisioned a one-milliwatt beam that could be focused onto a spot only one micrometer across, roughly twice the wavelength of visible light. Doing so, he said, would concentrate power to an impressive 100,000 watts per square centimeter in the much smaller spot. It was a grown-up physicist's version of concentrating the sun's energy with a magnifier, and he hoped it might produce some effects that would interest physicists. It was far short of what Gould had proposed to ARPA.

The big question was how to generate the population inversion necessary for stimulated emission. One possibility was zapping a gas with an electric discharge to excite the atoms, as Javan and Sanders had proposed. Schawlow said that Willard S. Boyle of Bell Labs and Gould also were working on other approaches using electric discharges. Townes had asked him to say something nice about Gould, so Schawlow made a point of citing one of Gould's ideas for a laser cavity.

But the heart of Schawlow's talk was on solid-state lasers. He thought one might be quite simple to build. "In essence, it would be just a rod with one end totally reflecting and the other end nearly so. The sides would be left clear to admit pumping radiation." Think of a straight glass rod, with its ends painted with mirror coatings of silver, so the light bounces back and forth between them. Pump light entering through the transparent sides of the rod could excite atoms in the glass. To get the light out of the cavity, one end might have a hole in the silver paint, so some light could emerge as a beam. The idea was simplicity itself. The problem was finding the right material.

Schawlow had studied ruby carefully, but it had disappointed him. He had given up on pink ruby, with low concentrations of the chromium atoms that colored the crystal. "The two strongest lines . . . go to the ground state, so they will always have some atoms in their lower state, and are not suitable for [optical] maser action," Schawlow wrote. It was a flat prediction of failure, a message Schawlow would repeat many times in the coming months.

Ruby had seemed promising at first, with bright red fluorescence lines highlighting a well-studied spectrum. Solid-state physics is so complex that physicists prefer to start with well-studied materials, because starting from scratch can add years of data collection to the quest for something suitable. But the closer Schawlow looked, the worse ruby seemed. The two brightest red wavelengths the crystal emitted came from chromium atoms dropping from an excited state to the lowest possible energy level, the ground state. This meant that producing a population inversion required exciting more atoms to the upper laser level than remained in the ground state. That was difficult but it was not necessarily impossible—if the chromium atoms absorbed and emitted light efficiently. Yet the amount of red fluorescence that Wieder had reported at the optical-pumping meeting and repeated at Shawanga Lodge implied that ruby was only about one percent efficient. That efficiency was so low that Schawlow didn't think he needed to bother trying to measure it directly. The measurements were difficult, and they didn't seem worthwhile after Wieder's experiment confirmed what Bell Labs colleagues had been telling him—fluorescence on the two red lines of pink ruby was inefficient.

Schawlow did hold out hope for red ruby, which contained at least 0.5 percent chromium atoms, because the two extra red transitions that appeared when chromium atoms were close together behaved differently than the transitions of pink ruby. The lower levels of those transitions are just a little bit above the ground state. At room temperature chromium atoms have enough energy to creep up from the ground state into the lower laser level, so the system behaves just like the two red lines of pink ruby. However, cooling the crystal close to absolute zero empties the lower laser level, so fewer chromium atoms would have to be excited to invert the population. This led Schawlow to hold out hope that cooling red ruby in liquid helium might open the door to a ruby laser. Yet the difficulties remained formidable, and Schawlow really wanted a better material.

A lively discussion followed Schawlow's talk, where Gould mentioned that pulsing a laser might generate pulses that at their peak might have power as high as a megawatt. It hinted at TRG's classified research program, but strictly speaking it couldn't be called a security leak because Gould still lacked a clearance. The idea of pulsing didn't stimulate Schawlow. His goal was a laser emitting a modest amount of light continually, the optical analog of the continuous solid-state microwave masers that were already running at Bell and elsewhere.

Much of the action was in the halls, at breaks, and in the evening. The conference was intended to encourage communications among scientists from around the globe working in several different fields. The organizers were particularly interested in the little delegation from the Soviet Union. (See figs. 8.1, 8.2.) Most

FIGURE 8.1. Microwave maser developers at Shawanga Lodge. From left, Jim Gordon, Nikolai Basov, Herbert Zeiger, Alexander Prokhorov, and Charles Townes. (From *How the Laser Happened: Adventures of a Scientist,* by Charles H. Townes, copyright 1999 by Oxford University Press, Inc. Used by permission of Oxford University Press, Inc.)

of them had never met Basov and Prokhorov, who showed up with a couple of scientific unknowns. It was a typical pattern in the Soviet era. Basov and Prokhorov were both well-connected members of the Communist Party, but the cautious Soviet bureaucracy always sent along a handler or two, typically quiet, burly types who spoke little English. American intelligence agents were rarely in evidence at scientific meetings, but they often grilled U.S. scientists after they spent time with Russians.

The silent presence of the handlers didn't stop the Russian scientists from an evening of vodka-lubricated partying. Laying out the liquor was part of the Cold War game, with each side hoping that someone on the other would drink so much that some crucial secret slipped from their lips. It was also a way to get acquainted. After a few too many drinks, Javan took Basov, Prokhorov, and a few others for a spin in his sports car, only to wind up stuck in the mud off the side of the road. The whole group got out and sank up to their knees in mud lifting the little car back onto the road.

Looking back, many attendees consider the conference as a turning point for the race to make the laser. Young scientists like Peter Sorokin of IBM were thrilled

FIGURE 8.2. Ali Javan (left) ant Nikolai Basov talk at Shawanga Lodge. (Courtesy of Dimitri Basov).

by the air of excitement. Many working on the laser returned home optimistic that success was within reach. Ted Maiman, invited to talk about his ruby microwave maser, came back wondering about the prospects of making an optical laser from ruby. He had sat and listened to Schawlow say that pink ruby couldn't work, but he wasn't convinced. He didn't think Schawlow had the whole story.

9

A DARK HORSE JOINS THE RACE

TED MAIMAN DIDN'T SEEM TO BE A CONTENDER in the laser race at the Shawanga Lodge meeting. He had been invited to give a paper, but it covered the compact liquid-nitrogen-cooled ruby microwave maser he had finished in July. It was an elegant device that represented important progress, but it was a microwave maser, and by September 1959, microwave masers were no longer on the cutting edge of physics. Microwave maser papers were part of the crowd at the conference; it was the handful of papers on lasers—officially "optical masers" at Columbia and Bell Labs—that stood out.

Masers had become "routine," the editor of the prestigious new journal *Physical Review Letters* proclaimed in an editorial published August 1, just weeks before the conference. Samuel Goudsmit had found his desk covered with papers on microwave masers. He had founded the journal to publish hot research in physics, but he felt that many maser papers "contain primarily advances of an applied or technical character and comparatively little physics." As a pure physicist, Goudsmit was not interested in mere devices or engineering. He decreed that maser papers could go elsewhere unless they contained "significant contributions to basic physics," and made it clear that few did.

That didn't mean no one cared about microwave masers. In fact, corporations and government agencies had far more money to spend on practical devices than on contributions to basic physics. The military generously supported research on microwave masers, and the avalanche of papers reaching Goudsmit's desk testified to widespread interest. But they represented incremental advances, or "practical" devices beneath the interest of pure physicists. Like reporters or hyperactive children, the research journal craved something new, the dramatic breakthroughs that are the stuff of headlines. Improving a microwave maser was a solid single, but it didn't bring home any runs on Goudsmit's scorecard.

Maiman had shifted from electronics to physics because he wanted bigger challenges than building just another microwave maser. With his contract to build the compact ruby microwave maser finished, his managers expected him to find a new project that would earn a new military contract. Until he did, Maiman's salary was paid from a pool of general research money that Hughes maintained. The cash came from overhead charges on military contracts that supported much of the work at the research labs. It was a way the Pentagon could encourage research in new areas not mature enough to justify a formal contract. Hughes Research Labs used the money to reward successful scientists, hoping their new ideas would lead to more contracts.

Maiman had begun thinking what to try next before the Shawanga Lodge meeting. Optically pumping a microwave maser seemed a logical and attractive next step. Maiman had used light in his dissertation research at Stanford, so he was familiar with optical measurements. He knew that optical pumping was a powerful and effective way to excite atoms or molecules to desired states. And he thought that optical pumping might give a microwave maser a very desirable property—a low level of background noise.

Every signal has some noise in the background, like the faint hiss of an audio cassette tape. Engineers want the level of noise to be as low as possible. Anything done to a signal can add noise—an increment called the "noise figure"—and this addition also should be as small as possible. It was particularly important to keep the noise figure low in microwave maser amplifiers, because their job was to increase the strength of very faint signals that could easily be overwhelmed by noise, including the faint emissions of distant radio galaxies, or the weak radar returns from distant objects. For masers, Maiman expected the noise figure to decrease if the frequency that excited the atoms was much higher than the frequency of stimulated emission. Light waves have a frequency about 10,000 times higher than the highest frequency microwaves, which looked like it would offer a big advantage.

Irwin Wieder was in the early stages of investigating optical pumping of a ruby microwave maser under a military contract at Westinghouse, and he had put Maiman on the mailing list for his regular progress reports. Maiman had his own ideas on the subject. He had learned much about ruby from developing his compact microwave masers, and it seemed a logical starting point for him to start studying the prospects for optical pumping of a microwave maser.

A second possibility lurked in the background, trying to make an optical laser. That posed a tougher challenge, but Maiman liked to apply his mind to tough challenges. He was well aware of the Schawlow–Townes paper and of Gould's proposal, and he knew Bell Labs, TRG, Columbia, and other groups were trying to make lasers. Yet before the Shawanga Lodge meeting he hadn't heard of much real progress. He made some preliminary calculations about the conditions needed to obtain oscillation on an optical transition, looking particularly at pink ruby, a material he knew well.

Schawlow and Townes had analyzed laser requirements in terms of the internal physics. They considered how long atoms stayed in excited states, the probabilities that atoms would absorb or emit light when making transitions between pairs of energy levels, and the nature of the resonant cavity. Maiman converted their formulas into engineering terms more useful in assessing whether a laser could oscillate. He balanced the gain in optical power each time the light made a round trip through the laser material against the amount of light lost on each pass, including energy absorbed by the mirrors and transmitted in the beam. If the increase in light power was larger than the losses, the total power would rise after each round-trip. The increase didn't have to be much. As long as the gain exceeded the loss by even a small fraction of one percent, the device crossed the threshold for oscillation. If the loss was higher than the gain, the laser wouldn't work. It's a practical technique still used in laser research and development.

At Shawanga Lodge, Maiman paid careful attention to the talks on optical lasers. Schawlow explained that pink ruby wouldn't work in a laser because the ground level of chromium was also the lower level of the potential laser transition. That meant that the only way to produce a population inversion on that transition was to excite more than half of the chromium atoms so they wound up in the upper laser level. Schawlow was convinced that was impossible. He held out some hope for a pair of transitions in red ruby that dropped to a state just slightly above the ground state. He thought that cooling red ruby close to absolute zero might reduce the population of the lower laser level almost to zero, allowing a population inversion and laser action on the two red ruby lines. Most of the audience left convinced Schawlow was right, and over the coming months he would repeat the argument many times.

Yet little things didn't seem right to Maiman. As he described the problems with ruby, Schawlow said, "You would not be able to deplete the ground-state population because the crystal would be bleached." Maiman thought the jargon of physics was hiding an error in logic. Bleaching normally means to make the color fade, but it has a more specific meaning in optical physics. Materials get their color from how they absorb or emit light. Bleaching a material means exciting all atoms out of a particular energy state, so they cannot absorb or emit light on transitions that start from that state. Maiman didn't see why bleaching would make it impossible to deplete the population of the ground state because bleaching *meant* depopulating it. If you bleached the ground state, you would invert the population.

Maiman's critical eye also noted that Schawlow had not shown specific calculations to prove his case against ruby. This is not uncommon at conferences, where speakers usually describe results rather than talk their way through detailed calculations. Yet Maiman wanted more than a hand-waving argument. He had a keen analytical mind, and he had studied under Lamb, a master of analysis. He suspected Schawlow's intuitive analysis had missed something, but he wasn't about to stand up and quarrel with an expert until he had done his own quantitative analysis and was sure of his position.

Irwin Wieder added to the bad news when he said that pink ruby emitted only about 0.1 watt of red light. Maiman knew Wieder well enough from their years at Stanford to feel comfortable taking him aside and grilling him on details of the ruby measurements. Wieder hadn't spelled out his estimate of ruby fluorescence efficiency at Shawanga Lodge, but from his answers Maiman could tell the number was low. His first inclination was to believe Wieder's results because they had worked together. Yet Maiman headed back to California with a nagging suspicion in the back of his mind that it might be possible to make ruby fluoresce more efficiently. Unless he could find some way to excite chromium atoms efficiently out of the ground state, he had no hope of making a laser from ruby, but Maiman came away convinced that the laser itself was a worthy pursuit.

That put him at odds with his immediate supervisor at Hughes, George Birnbaum, who like others came away from the Shawanga Lodge meeting discouraged about prospects for a ruby laser. Back in California, Birnbaum cited Schawlow's case against ruby as a laser material, and urged Maiman to try some something else. Maiman pushed to examine ruby more closely, and with a good track record at Hughes, he was allowed to go ahead.

If one thing was clear by September 1959, it was that the optical laser was not going to be an easy target. Nearly a year after Schawlow and Townes had put the idea into the public domain, no one had reported serious progress toward building a working laser. A vital ingredient was missing from the recipe—a material with the right characteristics.

The buzz at the meeting was that the ideal laser material should have four levels involved in the process of stimulated emission, as Javan had described for neon and Schawlow had suggested for red ruby. The starting point is the ground state, from which atoms are excited to a high energy level. The excited atoms quickly release part of their energy and drop to a lower energy level, and the nature of that lower level is crucial to making the laser work. Atoms stay only a short time in most excited energy levels, quickly releasing energy and dropping to a lower state. Like buckets with big holes in them, these states can't hold atoms long enough to sustain the steady population inversion needed for continuous laser action.

The hopes for four-level lasers depended on the existence of excited states which trapped atoms for much longer times. They are called "metastable" because it's hard for atoms in those states to drop to lower states. They're buckets with a few small holes, which a good hose can fill faster than the water leaks out. You can think of them as wide, sticky steps on the energy-level ladder, which keep the atoms in place until stimulated emission bumps them out of the upper laser level and down to the lower laser level. Ideally the atoms should drop quickly out of the lower laser level, to maintain a population inversion between it and the metastable upper laser level, as shown in figure 9.1.

Red ruby was a four-level system close to absolute zero, but not at room temperature where thermal energy gave chromium atoms enough energy to move freely between the ground state and the level just above it. Pink ruby lacked that extra energy level just above the ground state, making it a three-level system. In principle, it's possible to have a three-level laser where the atoms are excited directly to the metastable upper laser level, then drop to a lower laser level separate from the ground state. That should work as well as a four-level laser, but nobody had such a material in 1959. Ruby was a more difficult case, because the ground state was the lower laser level, and emptying the ground state looked very difficult.

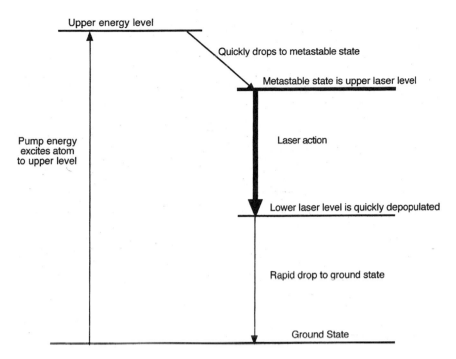

Upper energy level

Quickly drops to metastable state

Metastable state is upper laser level

Pump energy
excites atom
to upper level

Laser action

Lower laser level is quickly depopulated

Rapid drop to ground state

Ground State

FIGURE 9.1. A four-level laser can operate continuously as long as the upper laser level is metastable and the lower laser level is depopulated quickly.

The two leading ideas for making gas lasers were four-level systems, but neither appealed to Maiman. His analysis of the potassium metal vapor approach pursued by TRG and Columbia showed it had little hope of working, so he saw no point in trying it. At Stanford he had worked with gas discharges, and learned that energy transfer in them can be very complex because it can involve many possible energy levels. Maiman judged that making a gas discharge laser work would require endless months of patient experiments, adjusting everything from gas concentrations to the properties of the electric discharge until everything was just right for the laser to oscillate. That promised to take a long time, so he left gas discharges to Bell Labs and TRG.

Time and money were serious issues for Maiman. Bell Labs could afford to let its scientists play indefinitely on pet projects. Hughes couldn't. The company had some money for internal research, and Maiman had earned a share of it, but he knew he would have to demonstrate progress. He knew going for an optical laser was a big gamble, but his earlier successes had given him enough self-confidence to bet on his own judgement, and he could see the potential for a big payoff at the end. He thought his best chance lay with solids.

Solids offered a number of advantages. They pack many more atoms into a much smaller volume than a gas, and that higher density promised to make more excited atoms available to amplify light. The higher the amplification, the easier it should be to make an optical oscillator. Having worked with solids, Maiman considered them relatively simple, because atoms were fixed in place, and had only a few possible energy levels. That meant he wouldn't have to worry about the movement of gas atoms in all different directions, or the many energy levels possible for atoms moving about in a gas. Javan, Bennett, Townes, and Gould all had worked extensively with gases and felt more comfortable with them, but that was their privilege.

More importantly, solids promised operational simplicity. "The bottom line was that, in principle, a solid crystal laser could be designed to be very simple, compact and rugged," Maiman wrote. With a suitable solid crystal, he figured a laser would be ready to go—no need to adjust gas pressure and composition, run vacuum pumps, remove impurities from the gas, or fuss around with any of the other little things needed to keep a gas in the right state. Gas systems also required careful operation and maintenance. A metal-vapor laser was likely to require the tender loving care of a lab full of graduate students and technicians to keep it humming. A solid-state laser shouldn't. Transistors were already showing that solid-state meant reliable.

There's a vast difference between the operational simplicity Maiman sought in a solid-state laser and the theoretical simplicity that Townes, Schawlow, and Gould saw in the alkali-metal approach. Theoretical simplicity means a device that's simple to explain in textbook terms; it says nothing about how easy it is to build or operate. Potassium vapor is theoretically simple to describe because each atom has only a single electron in a valence orbit outside the cloud of inner-shell electrons. Yet it's a practical nightmare for experimenters because it requires extremely careful handling. The vapor must be heated to the right temperature, tends to condense in awkward places, and is dangerously reactive. An alkali-metal vapor laser would be at best the sort of sensitive physics experiment that kept Columbia graduate students in the laboratory for years.

A solid-state laser likely would be harder to describe in the textbooks, but Maiman didn't care. He wanted a laser that would be useful. That meant a solid-state laser that should operate at room temperature, not cooled with liquid helium or liquid nitrogen like the ruby microwave maser he had just finished. Cryogenics weren't quite as cumbersome as alkali-metal vapors, but Maiman had worked with them enough to want to avoid them. He wanted a practical laser, and that meant one operating at room temperature.

Designing it was a job that required a thorough understanding of both physics and engineering. Maiman had both, but he knew he faced a tough challenge. Looking back, he reflected in his autobiography, "I was a little brash. I would be thrusting myself, in a sense, into a technological Olympics. The competition was of the best quality and of international scope. But my competitive spirit won out. The challenge of working in the top league of such an exciting project, that had so many questions and problems to resolve, was very compelling to me."

If Maiman's ultimate goal was clear, the path to it was not. The prospects for making a laser depend strongly on the energy levels of atoms in the material, and the energy-level ladders of solids differ widely. Physicists can't calculate the energy levels of solids from simple textbook formulas. They have to painstakingly measure how the materials absorb and emit light at various wavelengths, and use data from those experiments to calculate how atoms jump between energy levels. That meant it was easiest to start with solids that had already been studied. Ruby was one candidate, but Schawlow and Wieder had cast doubts on its laser prospects. Another was gadolinium ethyl sulfate, a fragile crystal that Bell Labs had used in the first solid-state microwave maser. Maiman knew more about ruby, but he also considered the gadolinium crystal worth investigating.

The idea of studying gadolinium came from a new assistant Maiman had picked in June: Charles K. Asawa, a graduate student in spectroscopy at UCLA, who warned it would be difficult to make into a laser. Born of Japanese immigrant parents in 1920, Asawa grew up on the family truck farm in then-rural Norwalk, California. Physics and mathematics fascinated him early, but the realities of life got in the way. His father died when he was nine, and his older brother drowned in an accident four years later, leaving the 13-year-old Charlie as "the man of the family" to help his mother run the farm. He managed to take night classes in math and physics at a nearby community college, but after the Japanese attack on Pearl Harbor his family were among the 120,000 Japanese-Americans interned by the government. He, his mother, and three sisters spent the spring and summer of 1942 in a freshly paved horse stall at Santa Anita racetrack before being shipped in September to Rohwer Relocation Camp in Arkansas. The following year he was released to attend the University of Cincinnati, where he nearly finished requirements for a math degree before being drafted in 1944. Asawa volunteered for Japanese language school and became one of the 400 translators who went to occupied Japan after the war with General Douglas MacArthur. He earned his math degree in the Army, and after being discharged worked for several years before his wife encouraged him to start a graduate program in spectroscopy at UCLA. He was six years older than Maiman, but with jet-black hair he looked younger, so few at Hughes realized his age.

Maiman's first task for Asawa was to find the optical properties of ruby, particularly how it absorbed and emitted light, information vital to assess prospects for optical pumping of a ruby microwave maser—or a ruby laser. Asawa uncovered a pair of papers that described research by Japanese spectroscopists, but Maiman wanted a more detailed analysis of ruby, which was a tough job. Asawa was studying solid-state spectroscopy at UCLA, but he was working on a group of elements called rare earths, which interact relatively weakly with the crystals that contain them. Ruby was a much more difficult task, because chromium atoms interact strongly with the surrounding aluminum and oxygen atoms. Asawa gave Maiman his analysis in August.

The results were frustrating. Ruby met many basic requirements for a solid-state laser. The crystal was durable, and transparent enough for light from outside the crystal to reach the light-absorbing and -emitting chromium atoms inside. The chromium atoms glowed red when illuminated with bright light at shorter wavelengths, and the red glow passed through the transparent crystal. The spectroscopy looked promising. Yet two men who had studied ruby's properties, Wieder and Schawlow, had insisted that the process that made that red glow was very inefficient.

Maiman analyzed the ruby data, but he couldn't determine where the energy was leaking out of the ruby crystal, and that bothered him. If he could find where the energy was going, he hoped to find some vital clues to identifying what type of solid might work best as a laser. Perhaps he could even find a viable three-level laser, as an alternative to the more complex four-level schemes that Javan and Schawlow had pushed at Shawanga Lodge.

Solving the mystery of the missing energy required more careful optical measurements of ruby. Maiman asked Asawa to find a compact light source that could illuminate ruby with very intense light. Maiman also decided to buy an instrument called a monochromator to examine the wavelengths ruby absorbs and emits. Like a prism, a monochromator spreads out a spectrum from a bright source of white light, but it includes a narrow slit that slices out a very narrow range of wavelengths, to illuminate an object with a single pure color. It's a standard tool for spectroscopists, and Maiman had used one at Stanford, but Hughes didn't have one available. Maiman wanted a model that cost $1500, enough to buy a good used car at the time. It wouldn't have been a budget-buster at Bell Labs, but at Hughes it required the signature of department manager Harold Lyons, a level above Birnbaum. Lyons wasn't eager to spend the money, but Maiman finally persuaded him to buy the instrument.

The new instrument let Asawa and Maiman closely examine the spectroscopy of pink ruby. They poked and they probed, but they still couldn't find what happened to the missing energy. There was no optical counterpart of the puddle on the floor left by leaky plumbing. Suspecting something was wrong, Maiman decided to measure the fluorescence efficiency of ruby for himself. He didn't fully trust Wieder's experimental skills, and Schawlow had given little evidence to back up his claims for low efficiency. Even if they were right, Maiman hoped that tracking down the missing energy would give him new clues on how to proceed.

His course was still not set completely toward building a laser. The optically pumped microwave maser looked like an attractive intermediate step. Working on internal Hughes funds, he knew the project couldn't go on indefinitely. But Maiman knew ruby well enough to believe he could identify the discrepancy, and to hope that it might tell him something useful that would help him make a laser.

10

"EVERYBODY KNEW IT WAS GOING TO HAPPEN WITHIN MONTHS"— BELL LABS FEELS SAFELY IN THE LEAD

TED   MAIMAN   AND   HUGHES   RESEARCH   LABORATORIES
weren't on the radars at Bell Labs. Bell felt safely in the lead and confident of its staying power. Bell managers didn't skimp in the pursuit of promising ideas or haggle about spending $1500 on a potentially useful instrument. They sent 18 people to Shawanga Lodge, plus John Sanders, who listed himself as coming from Oxford University. That delegation was more than ten percent of the total attendance. Most returned optimistic that they were well on their way to a working laser. "Everybody knew it was going to happen within months," recalls Al Clogston, who was in charge of coordinating Bell's various laser projects.

The Shawanga Lodge meeting confirmed the impression that Bell was safely in the lead. TRG looked like the only challenge, with its million-dollar ARPA contract giving the little company resources to compete seriously. Gould had been full of questions at the meeting, but the organizers had invited only one other TRG physicist, Maurice Newstein. Gould had some friends at Bell. However, like Schawlow, Bell management discounted his laser project, and worried more that the TRG contract might push their optical maser research into the classified realm than about the chance that the upstart company might beat them in the laser race.

Top management gladly supported the quest for the optical maser, without worrying when its research investment would pay any dividends. Supporting research was a corporate tradition at AT&T, where top scientists and engineers like Rudi Kompfner climbed the ranks of research management and honed their skills at managing innovation. AT&T looked like a utility monopoly to the users of the hefty and nearly indestructible black telephones it made at its own factories, but behind the scenes it was a technology giant. The vast bulk of the work at Bell Labs was engineering, developing vital equipment the public never saw, like special-purpose electronic computers that automatically switched the growing volume of telephone calls around the country. The labs were working on video telephones, which its top engineers were convinced would be the next big thing. Basic research projects like the laser were the tip of the iceberg, highly visible investments in the long term. They had paid off handsomely before, and they kept Bell's cadre of top-level physicists from wandering off to universities.

Nobody at Bell expected it would be simple to make a laser. Sanders had spent his sabbatical at the labs demonstrating that a simple approach wouldn't work. The optimism came from Bell's rich resources, which were the envy of academic physicists, and an impressive depth of talent that allowed four separate teams to work on the idea after Sanders returned to England.

Schawlow had written off pink ruby, but he continued studying the spectrum of red ruby in a highly effective partnership with his technician George Devlin. Although Schawlow was an experimental physicist, he considered himself clumsy. He was more comfortable devising experiments and analyzing results than working in the laboratory. Devlin started at Bell with only a high-school education, little math, and no formal training in physics, but he had been a champion model airplane builder. He had a quick mind, a good observational eye, and the vital skill for understanding how things work. His skills complemented Schawlow's. Devlin was the one who had first noticed the two extra emission lines of red ruby when the two were studying samples cooled to near absolute zero. Schawlow looked at their measured wavelengths, 700.9 and 704.1 nanometers, and realized the extra lines came from the interaction of chromium atoms with each other that was possible at the high chromium densities of red ruby.

The two extra lines in red ruby were the only threads of hope Schawlow held for laser action in ruby, because their lower laser level was just above the ground state. They were slender threads. To keep thermal energy from exciting chromium atoms to the lower laser level, he calculated the crystal would have to be cooled to the temperature of liquid helium, four degrees Kelvin. He planned an experiment using the basic design he had described at Shawanga Lodge. White light would illuminate a rod of red ruby, exciting the chromium atoms to their metastable upper laser level. The ends of the rod would be polished smooth and parallel to each other, then coated with reflective films, so stimulated emission would bounce back and forth between them, each time passing through the rod and stimulating the emission of more light. One of the reflective end films would transmit a little light to form the beam.

Performing the experiment was a tricky proposition. The ruby rod had to be kept cold while it was illuminated with a bright light. That meant submerging the rod in liquid helium while still exposing it to light. Fortunately, Devlin unearthed a promising bit of hardware in a storage room full of surplus equipment. Liquefied gases are kept in insulated glass containers called dewars, which are essentially high-grade thermos bottles. Slipping a ruby rod into a normal dewar filled with liquid helium would keep it cold—but hide it away from any light source. What Devlin found was a specially made dewar that had a transparent finger of glass protruding from the side that could be filled with liquid helium. Slip a long thin crystal into the protruding finger, and the liquid helium could keep it cold during experiments. Another Bell physicist had used it for microwave experiments. Devlin figured it might work for an optical experiment.

A lamp emitting a steady light would quickly heat the rod above the desired temperature, so they turned to pulsed lamps. Devlin and Schawlow found a spring-shaped flashlamp big enough that the glass finger of the dewar could fit inside its couple of loops. The lamp had come from an instrument called a strobotac, which fired pulses of light at regular intervals to measure rotation rate, so it came with a power supply that generated electrical pulses precisely matched to the lamp's requirements. That avoided the problem of overloading lamps that McAvoy and Wieder encountered at Westinghouse. The lamp fired pulses of about 10 joules, which Schawlow thought should be enough.

Bell had plenty of ruby left from other experiments; Schawlow and Devlin found a ruby rod 2 millimeters thick and 50 to 60 millimeters long, about the size of a short ball-point pen refill. They had another Bell Labs specialist polish the ends to be smooth and parallel to each other, and apply a mirror coating.

Their plan was to slip the rod into the transparent finger, cool the rod by filling the container with liquid helium, and place the flashlamp around it. Schawlow hoped to see subtle but definite effects when the flashlamp was fired. Normally the white light from the lamp made ruby glow red with fluorescence emerging from both the sides and the ends of the rod. He expected stimulated emission would produce a different pattern. Stimulated emission along the length of the rod would bounce back and forth between the mirrors, but light emitted in other directions would leak out of the sides. That would make the light emerging from the ends much brighter than the light leaking from the sides. Schawlow also expected that stimulated emission would make chromium atoms drop from the excited state faster than normal fluorescence would, so the red glow should fade faster.

When the apparatus was ready, Schawlow went to a conference, leaving Devlin behind to do the experiment. Their plan was to turn up the power, so the lamp flashed brighter and brighter, and monitor light emitted by the ruby with sensitive instruments to see how it changed as the lamp power increased. That proved easier said than done. Each time the lamp flashed, it heated the rod, blowing liquid helium out of the transparent tip and spewing helium gas into the room. The liquid helium evaporated quickly from the dewar, with each fill lasting only about an hour.

More problems emerged as Devlin continued. The electrical pulses that fired the lamp induced stray electric currents into electronic instruments in the room. Stray light from the lamp spread around the room. The lamp was too short to spread its light evenly along the rod. And the curved glass covering the end of the rod bent light so much that Devlin wasn't sure that the red light he measured was coming through the end of the rod. Finally, after about five days, the rod broke at the end where it was clamped in place. Devlin decided the project was futile and tossed the rod into his desk drawer. "Almost every aspect of this experiment doomed it to failure," he recalled. The apparatus needed a major overhaul before he could hope to accomplish anything, and he had other work to do on superconductivity.

Schawlow didn't seem to have expected much and didn't bother going back to redesign the experiment. Too much remained unknown, and the prospects looked bleak. He wasn't sure how cold the red ruby would have to be to produce a population inversion and stimulated emission. He didn't know how much power he would need from the lamp. Schawlow saw no hope of success unless he completely redesigned the experiment. He decided that wasn't worthwhile and shifted his attention to other potential laser materials.

The choice reflected Schawlow's relatively low-key personality. He was not a competitive man, nor one who instinctively took charge of projects. He lacked the tightly-focused intensity that kept Townes working six days a week. He never gave Devlin a hint that they were in a race to make a laser. Schawlow wouldn't push to get costly new equipment to do new experiments. He sometimes blamed that reluctance on his training at the resource-poor University of Toronto, but it also came from deeper inside a man who had grown up with modest means during the Great Depression. Curiosity drove Art Schawlow, not the need to be the first or the fastest. He enjoyed listening to jazz, telling stories, and devising clever demonstrations. He told fellow Bell scientists that he wanted "to do physics," not to build devices.

The focus of Schawlow's curiosity was spectroscopy, the study of how materials absorb and emit electromagnetic waves, and understanding what that reveals about the inner workings of atoms and molecules. He and Townes had written the definitive textbook on microwave spectroscopy. Schawlow had switched to superconductivity at Bell, but he was happy to shift to the optical maser when Townes proposed the project. It brought Schawlow closer to his scientific roots: the vast, complex and elegant puzzle of spectroscopy. In his heart, Schawlow was a spectroscopist rather than a device builder. Giving up on ruby meant he had to examine other potential laser materials, which was a perfect excuse to go back to spectroscopy.

His goal was to find a solid that would behave as he wanted ruby to behave. The material should be transparent, so light could pass through it, but contain some atoms that absorbed and emitted light. Bell's experts in crystal growth went to work adding dashes of light-emitting elements to crystals that were otherwise transparent. One added rare earth elements to transparent crystals called garnets, best known in the everyday world as the hard grit on some types of sandpaper. Others added chromium to magnesium oxide and to gallium oxide. Schawlow studied their spectra carefully, but none of them looked especially promising as laser candidates.

Lacking a burning desire to build a device, Schawlow was easily distracted from the laser hunt by the outside world as well as by spectroscopy. A popular speaker, he accepted many invitations to give talks. He accepted an invitation from Townes to pinch-hit at Columbia as a visiting professor in the spring semester in 1960, getting a chance to try the academic life. Schawlow enjoyed his work at Bell, but didn't put in long hours, rarely showing up at the lab on evenings or weekends. "It was sort of like a job," he recalled later, not as consuming as being a university professor.

With Schawlow's progress stalled, Bell's best bet seemed to be Ali Javan and Bill Bennett, who were putting in long hours in a lab across the hall from Schawlow's.

Management had welcomed Javan's suggestion to hire Bennett. While they liked Javan's ideas, they thought Bennett, who had become an expert on energy transfer between gas atoms, would be better in the crucial experimentation steps. Bennett had arrived in early summer, but his collaboration with Javan had gotten off to a bumpy start. He had thought he was to be an equal partner and was disturbed to find that Javan had published the first description of his ideas for a gas discharge laser under his name alone. Yet Bennett had little choice but to go ahead with the project; he had three small children and had already resigned his job at Yale. He initially stayed in the background and wasn't invited to the Shawanga Lodge meeting. Javan became the point man, representing the project to the outside world.

Javan's tremendous energy went into overdrive as he focused intensely on the laser quest. The thrill of the chase soon soothed Bennett's ruffled feathers, and he matched Javan's pace. The two thought they had caught something big, and soon they were working up to 90 hours a week, virtually living in the lab. That was unusual at Bell; most scientists worked 40-hour weeks like Schawlow, and the place was virtually deserted by 5:30 P.M. Herriott helped with the optics when needed, but didn't match Bennett and Javan's long hours.

Bennett and Javan were perfectionists, trying to measure every possible relevant parameter before they set up their experiments. It came from the rigorous approach they had learned at Columbia, and it could be a valuable trait in tackling tough problems. Jim Gordon had needed years to get the first microwave maser operating, and the problems posed by the optical laser looked even more formidable. Just as Gordon and Townes had blazed their own trail to the microwave maser, Bennett and Javan needed to blaze the road to the laser. No one knew the best mixture of helium and neon to put inside a laser tube, the right way to pass an electric current through the gas, the right way to measure the laser gain inside the gas, or the right design for a resonant cavity.

They were exploring new territory, and that opened up a host of problems that had to be solved before they could actually try to make a laser. Bennett considered three problems to be crucial. One was establishing that extra energy absorbed by helium atoms was transferred to the proper excited states of the neon atoms. Second was measuring the lifetimes of both high and low energy levels of neon, to tell if it would be possible for them to produce a population inversion. The third was to show that stimulated emission actually would amplify light passing through the excited gas, producing measurable gain. All those conditions had to be met before they felt ready to try making a laser resonator.

The experiments needed to solve those problems were not simple or straightforward. The neon wavelengths where Javan expected stimulated emission to be the strongest were longer than about one micrometer. The human eye couldn't see that wavelength, and the best light detectors available at the time responded only very weakly to it. They had to conduct some experiments at liquid nitrogen temperature, 77 degrees above absolute zero, because the thermal energy present at room temperature generated too much noise. And Javan and Bennett had to develop entirely new techniques for some measurements that no one else had tried before. All that took time.

Bennett had measured energy transfer between gas atoms before, and was able to adapt those techniques for measurements on helium and neon. The difference in energy between helium and neon levels was fairly small, so he expected the energy to be transferred fairly efficiently between the two. The measurements showed an efficiency only about one-tenth of what he had hoped for, but that didn't seem like a fatal problem.

The prospects for making a laser did depend critically on how long neon atoms stayed in the upper and lower energy levels involved in a transition. The atoms had to stay a long time in the upper state, so it could accumulate a large population. They had to drop quickly out of the lower state, so it contained only a very low population. You can think of each energy level as a leaky bucket with water pouring in from above. The upper level should be a long-lived metastable state, like a bucket with only a few tiny holes, so the incoming water fills the bucket faster than water dribbles out the bottom. The lower level should be short-lived, like a bucket with big holes that let the water pour through the bottom as fast as it enters the top.

Lifetimes aren't easy to measure directly, but Javan and Bennett were able to estimate lifetimes using a formula that related them to other quantities that were easier to measure. The results were encouraging, but the uncertainties were large, and Bennett spotted some new Russian research that indicated the atoms probably stayed too long in the lower state of neon for their purposes. Worried that such a long lifetime might make a laser impossible, Bennett decided to try direct measurements of the lifetime to see if the Russians were wrong. He set up a complex apparatus that fired electrons into neon and timed how long the excited atoms took to emit light after they absorbed energy from the electrons. Unfortunately, he found the process generated a lot of noise, so he would need to collect many hours of data and feed it into a complex electronic analyzer to extract meaningful results. He knew exactly the instrument he needed, which had just come on the market for nuclear physics research. He also knew it cost $26,000, a sum he thought was astronomical. He had little hope of getting something so expensive, but nonetheless asked Bell's manager of physical research, Sid Millman, about the new analyzer—and was astonished when Millman approved the purchase on the spot. Over the weekend they bought it and hired a light plane to bring it from the manufacturer in New Haven to the Murray Hill, New Jersey laboratory. With that key instrument, Bennett and Javan found that neon atoms stayed at least five times longer in the upper state than in the lower one, indicating gain should be possible.

Bell was investing generously in equipment for Javan and Bennett's lab. The giant corporation had deep pockets, and its research managers thought Javan and Bennett were their best bet in the laser race. Yet managers also encouraged other Bell scientists to work on lasers. The company could afford it, and there was no guarantee any obvious approach would succeed. Like ARPA, Bell was hedging its bets by looking at a variety of approaches to the critical problem of producing stimulated emission.

Semiconductors were natural candidates. Physicists were exploring how adding small amounts of impurities affected the electronic properties of silicon. Silicon transistors were replacing the first generation of transistors, made of less durable germanium. Other semiconductors had been made to emit light, producing the first light-emitting diodes or LEDs. At first glance, producing stimulated emission seemed like it might be a logical next step. Yet formidable problems remained to be overcome. Semiconductor physicists were still working out the nature of energy levels in semiconductors. They knew them well enough to make transistors, but they didn't know how well they could hold energy. Nor did they know how to get the states they produced to release energy as light rather than in some other form. Transistors worked by controlling the flow of electrons through the semiconductor. LEDs generated light when an electron moving through the crystal dropped into a vacancy where an electron was missing, releasing its extra energy by spontaneous emission. Yet at the time, the best LEDs released only a tiny fraction—about 0.001%—of the energy as light. The best-developed semiconductors, germanium and silicon, didn't release any light at all.

Willard Boyle, a Canadian physicist who managed a small group at Bell, thought he had a way around those problem, and with David Thomas filed a patent application on a way to make a semiconductor laser. They thought that a bright pulse of light could excite electrons so the electrons broke free of their atomic bonds and moved about freely in the semiconductor crystal. Each negatively charged electron would leave behind a positively charged "hole" which also could move about the crystal as electrons jumped from atom to atom, leaving other holes behind. Boyle hoped the electron-hole pairs could retain their energy long enough to produce stimulated emission.

It was a reasonable idea given the current knowledge of semiconductors, and Boyle enlisted Don Nelson, a young physicist fresh from the University of Michigan, to work on it. They started with silicon, hoping they could manipulate its properties so it released some energy as light. When that didn't work, they tried gallium phosphide, a semiconductor that normally did release energy as light. That didn't solve the problem, and after several months they gave up when they realized that all the energy went into moving electrons and holes, not into light.

Other Bell physicists were working on a variation of Schawlow's plan to excite atoms in transparent solids with light from an external source. Geoffrey Garrett and Wolfgang Kaiser started with calcium fluoride, a transparent crystal that was fairly well known at the time. Bell had specialists in crystal growth on their staff, so Garrett and Kaiser had them grow calcium fluoride crystals containing small amounts of other elements. They doped the crystals with atoms that they expected to act like the chromium atoms in ruby, absorbing white light from an external source and emitting light at longer visible wavelengths. Their goal was to find a set of energy levels that would fit the ideal formula for a laser, so exciting them with white light would produce a population inversion between a metastable upper level and a lower level above the ground state.

They concentrated on the family of elements called rare earths, which Schawlow was also considering. Relatively uncommon, the rare earths are very similar to each other chemically because they usually have three outer-valence electrons. Spectroscopists had spent considerable time studying rare earths, and had collected large amounts of information on their properties. Rare earths also don't integrate themselves as tightly into the structure of host crystals as the chromium atoms do in ruby, so their behavior was relatively well understood and didn't vary greatly among host crystals. Those features made them a good starting point, and rare-earth solid-state lasers eventually would become important.

Garrett and Kaiser sought to make a laser from a crystalline rod, with its ends coated to reflect light back into the rod. The coatings would allow a small fraction of the light to leak out in the beam. On paper their ideas looked good, but progress was slow. It's hard to grow crystals of the high optical quality they needed for their experiments. When they did get good crystals, they found that the atoms they added to the calcium fluoride absorbed light at too narrow a range of wavelengths to collect much energy.

Two other Bell researchers attacked an important theoretical problem—predicting the exact patterns that stimulated emission would form as the light bounced back and forth between a pair of mirrors at the ends of a cylinder. Gould had the intuitive sense the light would emerge from a partially transparent mirror as a beam, but intuition didn't satisfy Bell's rigorous physicists. They had pushed Schawlow and Townes to refine their original ideas on cavity resonance before publishing their paper, but they had not been satisfied with the results. When the subject came up at an internal seminar, Javan and Bennett suggested the light would form a ring-shaped pattern on the surface of the partly transparent output mirror. The specialists in resonances were still not satisfied. They wanted a solid quantitative model to predict what they should see when and if someone got a laser working.

Rudi Kompfner asked Gardner Fox and Tingye Li to analyze the optical resonance more thoroughly. They started by thinking of the mirrors as a pair of antennas sending signals back and forth to each other. That let them draw on a base of information engineers had built developing antennas to transmit radio waves. The basic physics is similar; light waves and radio waves are both forms of electromagnetic radiation. What differs is the scale; light waves are 100,000 to a million times shorter.

Resonances occur when a whole number of waves fit exactly inside a cavity. As the resonant waves bounce back and forth, their intensity builds because their peaks and valleys are always in the same place. Many resonances are possible, each of which is called a mode. The principles are the same for sound waves, microwaves and light, with the modes depending on the size and shape of the cavity and the length of the waves. Calculating the possible modes is fairly simple for microwaves because no more than a few waves fit inside a microwave cavity. Laser resonances are more complex because thousands or millions of light waves can fit between the two mirrors on opposite ends. Fox and Li decided the best way to model the complex resonance was to calculate how the light waves behaved on a computer.

Computer models are standard today, but the field was young in 1959. They started work on an IBM 650, a room-sized monster filled with vacuum tubes, and a central processor that literally weighed a ton but was less powerful than a modern pocket calculator. They soon stepped up to a more compact computer that used transistors but was not dramatically more powerful. They couldn't rely on existing software; they had to write their own programs in Fortran, the first computer language developed for scientific applications. The programs had to be punched onto decks of cards, which were hauled from the Holmdel lab where Fox and Li worked to the main physics lab in Murray Hill. Writing and debugging programs that way was a slow and tedious process. If all went well, the program ran for two to three hours at a time at night, and the next day they received their cards back with a pile of printed output to analyze. If something went wrong, even a single mispunched letter in one card, they got the deck back with a printout containing only cryptic error messages that they had to decode before submitting a corrected deck for another run the next night.

Bell researchers held internal seminars to discuss their work and new ideas, but didn't say much to the outside world after the Shawanga Lodge meeting. Although they were making progress, their advances were too modest to merit trying to publish a new paper. The editor of *Physical Review Letters*, Sam Goudsmit, echoed the feelings of other journal editors when he wrote an editorial complaining that some physicists were padding their resumes by reporting each minor bit of progress in many separate short papers. Goudsmit vowed he wouldn't put up with that and would reject any papers that reported only incremental progress. Javan and Bennett planned to wait until they could conclusively demonstrate a working laser.

Everyone knew that others outside Bell were also working on lasers, so Bell scientists tended to be coy in responding to questions about their work. Most outsiders were cautious as well. Only the little group at Columbia was open about their activities, and they were not making much progress with potassium.

11

A CRASH PROGRAM
AT "PIPSQUEAK INC."

THE ONLY SERIOUS COMPETITION BELL SAW on its horizon was TRG. The two had some obvious similarities. Both had plenty of money to spend on laser research, Bell from the parent AT&T, and TRG from the Pentagon. Their staffs of bright young physicists were heavily laden with men who had gone through Columbia's physics department. The two groups knew each other. TRG had hired Ben Senitzky and Ron Martin away from Bell, and Bennett had consulted for TRG. Yet there were also strong contrasts.

Bell had a sterling track record of excellence, with financial resources and in-depth brainpower far beyond those that TRG could command. In the glory days, Bell's top scientists worked largely on projects of their own choosing. Managers like Rudi Kompfner spread ideas freely among members of the technical staff, creating an environment that encouraged innovation. Professors spent their sabbaticals at Bell, enjoying the luxury of laboratories far beyond the means of their universities. Top young physicists came to Bell after earning their doctorates. Some saw a few years at the labs as a stepping stone on their way to a professorship. Others liked the place so much that they settled in, hoping to spend their whole career in a laboratory without bored undergraduates or faculty committee meetings. It was a comfortable environment where ideas hummed in the air, and many people worked at their own speed on projects that interested them.

TRG was just a few years old, with only a few dozen employees, making it swifter and far less bureaucratic than the giant Bell Labs. Gould worked directly with TRG president and cofounder Larry Goldmuntz, while layers of management insulated Javan and Schawlow far from the top of the giant AT&T. At Bell, the laser was just another research project—although clearly a promising one. The laser project was one of TRG's biggest contracts, making it a top priority. TRG scientists didn't have the luxury of pursuing their own interests on company time; they were paid to work on contracts, or to generate ideas for new ones. TRG had to satisfy military priorities and security requirements that had no impact on Bell.

From the viewpoint of the 1950s establishment, the mighty Bell Labs seemed to have a clear advantage. Bell tended to look down at TRG as "Pipsqueak Inc." Yet looking back from a modern mindset, TRG's small size and corporate agility would make it seem a better engine of innovation than the corporate behemoth.

The reality was far more complex. Bell researchers were pursuing their own ideas largely on their own timetables. Gould was the font of TRG's ideas, but his lack of a security clearance gummed up the organizational works. The company had to put Daly in charge of the classified project because he had a high-level clearance as well as being a seasoned manager.

Paul Adams of ARPA felt Gould's clearance was vital for the laser project to succeed, and he promised to work on it. Yet his clout didn't reach into the military security system. The weight of pushing the clearance through fell on Yarmolinsky and Leventhal, lawyers who knew the ropes in Washington. They spent hours talking with Gould, and more hours interviewing his family, friends, and coworkers. Most were cooperative, with the notable exception of his ex-wife, who—steadfast in her leftist politics—had decided Gould was an enemy of the people and would do nothing to help him. The task facing them was a formidable one.

The lawyers' job was to present Gould's past indiscretions in the most favorable light, and establish him as a loyal citizen who wanted to help the government. There was no mistaking his interest in capitalism. Gould sounded like an entrepreneur when he talked about wanting to earn a million dollars from his inventions. Gould insisted he had broken with the Communist Party by 1950, disillusioned by the 1948 Soviet takeover of Czechoslovakia. But that wasn't enough for a former communist to pass the government's loyalty test. They wanted him to turn against his former friends and name names.

Gould wouldn't do it earlier, and despite the high stakes of his laser patent, he wouldn't do it now. He was no communist, but the security establishment often made little distinction among liberals, socialists, and leftists. The system had cast him as a political dissident, and he accepted the dissident's code of not turning his fellow dissidents over to the authorities. His stubborn stand was a triumph of principle over a tendency to avoid touchy issues and confrontation.

The lawyers spent long hours with him, extracting views they digested into a first-person explanation of his actions for submission with his clearance application. The lawyers worried about shades of meaning, but Gould couldn't remember key incidents. They did their best to craft a statement that would avoid offending the hard-line anti-Communist slant of the security establishment without seeming evasive or untrue. They wrote that Gould had not lightly applied for a clearance. He had given up applying for a passport in 1956 because the government was denying passports to leftists, and he didn't expect it to be approved. He had turned down an earlier job offer that would have required a clearance. This time was different, and not merely because he had a financial stake in the result, said the statement filed with his application in July 1959. "What tips the scales now is the belief that I have an important and perhaps unique contribution to make to the safety and security of the country." He cited the prospects for advances in missile detection and interception that had excited ARPA.

Leventhal began to feel optimistic when he began talking with Pentagon security officials after filing the application, but that feeling was soon dashed. One of Gould's former friends, Herbert J. Sandberg, tried to paper-over his own security problems by talking to the FBI about Gould. Senator Joseph McCarthy was dead, but lies and innuendo were not. The government didn't need hard information and verifiable facts to deny a security clearance. All they needed were statements, claims, and suspicions. Federal agents pressed scientists with security troubles of their own to prove their loyalty by implicating others. Sandberg had picked Gould.

By August, Goldmuntz was worrying about the mounting legal bills. By the fall, Gould himself was growing impatient with the unresolved case. The Shawanga Lodge meeting had come and gone, making the growing interest in lasers abundantly clear. Gould was anxious to get to work, but he couldn't until the government cleared him, and the security bureaucracy moved slowly indeed when their suspicions were aroused.

Gould's security woes complicated the laser project. Scientists applying for work interviewed with Gould, who radiated enthusiasm, and gave the impression he was in charge. Yet when they arrived they found themselves working for

Daly. Most of them were quite young, full of energy and enthusiasm, but lacking much experience relevant to the task of building a laser, so they would have benefited from Gould's guidance. Yet Gould was relegated to spending much of his time working on TRG's unclassified projects.

His isolation was physical as well as organizational. TRG has settled into a standard-issue brick industrial building at Two Aerial Way in Syosset, Long Island, New York, which served as the corporate headquarters and main laboratory. The laser project required more space, so TRG rented an identical building next door at Four Aerial Way, where they laid out offices and labs to meet government security requirements. Gould remained in a windowless office in number Two, outside the zone reserved for classified work. Daly moved the laser group into number Four, where it eventually grew to about 20 people.

As the man who had written the patent application and the laser proposal, Gould was the project guru. Members of the laser team had to ask him questions carefully so they didn't reveal what they were doing. Daly learned to dance around security issues by phrasing his questions with great care: "If we were working on a laser, which I'm not saying we are, and it wouldn't lase, what would we do about it?" Gould could suggest experiments for the other researchers to perform, but they couldn't come back and tell him the results—or ask him to interpret them.

It would have been a frustrating position for anyone, and it was particularly difficult for Gould. He was competent at physical calculations and theoretical analysis, but his gift was an intuitive understanding of how things work. He worked best in a hands-on environment, where he could see how a system behaved, then adjust it to see what happened. The security restrictions left him virtually handcuffed and blindfolded, without the feedback he needed to refine his ideas. Friction inevitably grew between the frustrated Gould and the strong-willed Daly. Each thought he should be running the laser project.

The situation slowly became a surreal farce that resembled Catch-22, the novel Joseph Heller was writing in New York, as security red tape snarled Gould and TRG. Federal agents classified Gould's notebooks, then diligently confiscated them from him because he wasn't cleared to see his own work. No longer naive in the ways of security, Gould saw that one coming and made clandestine copies for his personal use.

Gould had suggested a host of avenues to explore for making a laser, and the challenge for Daly and Goldmuntz was to concentrate TRG's efforts on the approaches that looked most likely to work. The program concentrated mainly on three fundamental approaches to generating stimulated emission. One was the optically pumped potassium vapor scheme, an example cited both by Gould in his patent application and by Schawlow and Townes in their paper. The second was exciting atoms in a solid with light from an external source, as Gould had proposed in his application, and Schawlow had investigated with ruby. The third was passing an electric current through a gas-filled tube to excite the gas by collisions of the second kind, the same technique Javan had proposed for helium and neon.

Goldmuntz thought optically pumped potassium vapor was "the quickest thing we could reduce to practice, in the cleanest way," so he gave it top priority. TRG's calculations indicated that a potassium laser was not likely to be powerful enough to be useful for any kind of weapon. However, the physics looked simple, and Goldmuntz thought the medium would be "pretty controllable," making it a reasonable starting point for experiments. Independent analyses by Gould and by Schawlow and Townes had shown that pump light from a potassium lamp was a good match to absorption lines of potassium vapor, and should produce a population inversion. It looked attractive as a first step, to prove that the laser concept was feasible before moving on to a larger-scale program to develop more powerful and practical lasers. TRG made its living from research contracts, and a successful proof of principle experiment was an ideal way to land more research contracts to make bigger and better lasers.

TRG assigned the potassium vapor project to Paul Rabinowitz, a recent hire who had been working for a pittance at Brooklyn College to pay the bills while he did graduate work in physics at New York University. He decided he needed a real job when his wife got pregnant, but interviews with several companies initially generated no good prospects. TRG left him dangling after a 1958 interview, but called him back for a second interview early the next year. After an interview with Gould, Daly, and two other TRG physicists, Rabinowitz landed a job at what seemed the princely salary of $6500. TRG initially put Rabinowitz to work developing a microwave maser for General Electric based on another idea of Gould's, optically pumping rubidium atoms to excite them to produce stimulated microwave emission. When the laser contract came in, it was a logical step to assign him to work on optically pumping potassium for a laser.

Although potassium vapor looked most promising to Goldmuntz for a demonstration, the Pentagon considered it the least sensitive of TRG's projects because it had been discussed openly and promised only limited power. Nonetheless, it was officially classified, and Rabinowitz needed a security clearance to work on it. He could ask Gould questions, and Gould could suggest experiments to conduct. However, Rabinowitz couldn't tell Gould anything about the results, or invite Gould to assist him in his secure laboratory.

The experiments required elaborate apparatus to vaporize the potassium, seal the reactive metal safely from the atmosphere, and excite the vapor atoms. The hot metal vapor condensed out on the sealed glass tubing, coating it and reacting with trace impurities. Rabinowitz had made only a little headway when he took a 4000-volt electric shock in a laboratory mishap and was sent to the hospital overnight for observation.

On the way out the door, Rabinowitz grabbed a physics textbook to keep his mind busy. Browsing through it, he found that atoms of cesium, a closely related metal, could be excited with ultraviolet light at a wavelength of 388.8 nanometers emitted by helium. Gould had worked with cesium beams earlier, and he had proposed optically pumping cesium as well as potassium. A big advantage of the cesium scheme was that the light came from a different material than they were trying to excite, which offered a way around problems they were suffering with potassium. "It was clear that if you could excite this state, you would have an inversion somewhere; whether you would have enough was the question," Rabinowitz recalls.

Meanwhile Gould went shopping for more scientists at an American Physical Society meeting and found Steve Jacobs, a spectroscopist with a doctorate from Johns Hopkins. After five years at Perkin-Elmer, a Connecticut optics company, he fell under Gould's spell and came to TRG to work with Rabinowitz. When Rabinowitz was back on his feet, the two proposed switching to cesium. Daly initially wasn't sympathetic, but changed his tune in a few weeks after Oliver Heavens, who was filling in for Townes at Columbia, mentioned that Abella and Cummins were switching to cesium when he gave a talk at TRG.

Rabinowitz and Jacobs talked with Gould about cesium, and he helped them design a simple experiment to check that cesium would work. They put a helium lamp next to a small sphere containing hot cesium vapor. When they switched the helium lamp on, the cesium atoms absorbed some of its ultraviolet light, then fluoresced at visible wavelengths as they released the absorbed energy. It was a simple and encouraging result—but it was classified, so they couldn't show Gould.

Daly took more interest in building an optically pumped solid-state laser. He put the effort under Ron Martin, hired away from Bell Labs, a senior physicist who was older than most of the young physicists on the laser project. Martin had worked with Derrick Scovil, the Bell physicist who built the first three-level solid-state microwave maser. The two had looked at the prospects for making a version of the three-level solid-state maser at optical wavelengths, and recognized that a key issue was finding the right transition for stimulated emission. They decided that pink ruby didn't look good because the ground state was the lower laser level.

At TRG, Martin began looking for solids suitable for optical lasers. Part of the job was assessing the prospects for using well-known materials, but he also worked on growing new types of crystals that he expected to have better properties.

One possibility was transparent crystals that contained small amounts of rare-earth elements such as neodymium. Those elements had attractive transitions, which released light when the rare-earth atoms dropped to energy levels that were well above the ground state. With that energy-level arrangement, they hoped that exciting only a small fraction of the rare-earth atoms to the upper laser level could produce a population inversion because the lower level should be empty.

Martin's group devised an elaborate scheme to grow crystals of calcium tungstate that contained low concentrations of rare-earth elements. They expected light to pass through the calcium tungstate to excite the rare-earth atoms, producing a population inversion. The material looked good on paper, but the crystals proved dauntingly difficult to grow. They succeeded in making crystals that were transparent, but the crystals were not optically perfect. Flaws inside the crystals bent the light passing through them. That was a serious problem because light had to pass straight through the laser crystal as it bounced back and forth between the mirrors on the two ends. Progress on the solid-state laser stalled as Martin's group tried to tame the crystal growth problem.

Ruby got a second look because the crystals were readily available and much easier to grow than calcium tungstate. Yet even ruby crystals were far from optical perfection, and the imperfections made experiments difficult. Daly knew of Schawlow's ruby experiments, so he invited Schawlow to visit TRG and describe his work. When he talked to the TRG physicists, Schawlow threw a big dose of cold water on the idea of making lasers from pink ruby. Schawlow still held hope that dark ruby could be used in lasers, but TRG soon found problems there as well. Martin calculated that a steady light source powerful enough to produce a population inversion in ruby would also melt the crystal. That left the possibility of making a pulsed solid-state laser, which didn't seem to excite anyone at

TRG. Convinced that they needed new materials before they could try building a solid-state laser, they turned to research on crystals.

Gould's patent application proposed several combinations of elements that might be used to make a laser using collisions of the second kind. Firing an electric current into a gas promised a way to deliver much more energy into the gas than was possible with optical pumping. The challenge was finding a way to extract the energy in the form of a laser beam. That depended on finding the right combination of two gases, one that could absorb the electron energy and efficiently transfer it to a second gas to produce a population inversion. Ali Javan had settled on helium and neon as the best prospects. Gould had listed helium and neon along with several other gas combinations—among them mercury and thallium, argon and krypton, and xenon and mercury. However, he thought the best prospect was a combination of the rare gas krypton and mercury vapor.

The idea was for krypton to absorb energy from electrons passing through the gas, exciting the krypton atoms into a state that could transfer their extra energy to the mercury atoms, producing a population inversion in mercury. Mercury was a logical choice; its spectrum and energy levels were well known. With the right kind of excitation mercury generates bright visible and ultraviolet light. The ultraviolet emission excites phosphors coated onto fluorescent tubes; the visible lines make mercury vapor lamps bright. Gould predicted discharge-powered krypton-mercury lasers might generate beams of light powerful enough to be weapons. That prospect interested the military so much that they made work on discharge-powered lasers one of the most highly classified parts of the TRG laser project—making sure that Gould would remain in the dark without a clearance.

Gould recruited Ben Senitzky (shown in fig. 11.1) to run TRG's experiments on discharge-powered lasers. He came from Bell Labs, where he had worked a few years after earning his Ph.D. under Rabi at Columbia. At TRG he was the experimentalist working with three theoretical physicists, who analyzed the issues involved in making the krypton-mercury scheme work as a laser. Like Bennett and Javan, he faced a tough set of experimental problems. Senitzky needed to measure the precise light intensity across a wide range of wavelengths, so he built an extremely sensitive instrument called a spectrophotometer that told him how intense the light was at each wavelength. That was vital information for identifying the transitions in any material, and for diagnosing what was happening in the gas.

FIGURE 11.1. Gordon Gould (left) and Ben Senitzky with a millimeter-wave amplifier, one of the unclassified projects Gould pursued at TRG while waiting for his clearance. (Courtesy of Gordon Gould)

In the krypton mercury mixture, he wanted to check if the energy was being transferred from the electrons to the krypton atoms and then to the mercury atoms, as predicted. If all those stages worked right, that energy transfer should invert the population on the mercury transition, producing stimulated emission that amplified light on that transition. The theory looked good, but the measurements didn't show the gain they wanted to see. "We came very close to an inversion, but we could never say that we had actual amplification," Senitzky recalled.

TRG had come up against a problem shared by everyone in the laser race. Amplification is much harder to measure than gain, although it was far from obvious at the time. To a physicist looking at the physical processes involved, amplification was a step on the way to oscillation. Physicists tend to take one step at a time, make sure it works, then go on to the next. It seemed logical to check if the light was being amplified before adding the mirrors to try to make an oscillator.

Yet that logic had a fatal flaw, because a laser can oscillate if it has a very small amplification. As long as the gain is larger than the losses, stimulated emission will build up the light intensity as it bounces back and forth. It is very hard to measure a one percent increase in the power of amplified light, but in an oscillator the beam bounced back and forth through the laser cavity thousands or millions of times, and those tiny increases built up to a much stronger signal that could be measured easily. An electronic engineer used to working with electronic oscillators might have spotted the problem, but it wasn't obvious to TRG physicists as they struggled to make a working laser.

TRG also was looking at another key part of the laser puzzle, designing an optical resonator. Gould and Schawlow had separately proposed making an oscillator by placing flat mirrors parallel to each other on opposite ends of a laser, to form a Fabry–Perot resonator. It was a simple and elegant design, but it suffered an important practical problem. It's very hard to align two flat mirrors precisely enough for light to oscillate between them if they're a couple of feet apart—unless you have a source of coherent light so you can compare the positions of the mirrors on the scale of a wavelength of light. In short, you essentially needed a laser to make a laser by aligning a pair of flat mirrors.

Gould recognized the problem, and his patent application proposed alternative designs that replaced flat mirrors with more complex optical devices that would automatically align themselves so light would bounce back and forth between them. One approach was to use optical devices called retroreflectors, often called corner cubes because they are essentially the corner of a cube, with three flat mirror surfaces meeting at right angles. If a light ray enters a corner cube, it bounces off the flat surfaces and emerges headed along the same line that arrived on, but in precisely the opposite direction. It was an elegant approach to avoiding the problem of mirror alignment; just put the retroreflectors on the ends of the laser, and the light would bounce straight back and forth between them. In fact, the scheme could work with a single retroreflector and a flat mirror on opposite ends of the laser. Gould figured he could achieve the same automatic alignment effect by using a special type of prism called a roof prism on each end of the laser.

The problem with Gould's idea was that it was difficult to make three flat mirrors and align then precisely perpendicular to each other in a retroreflector. The roof prisms proved simpler, but it still took time to make them, and Jacobs and Rabinowitz had plenty of other work to do.

They also considered another approach to automatic cavity alignment. A few years earlier French astronomer Pierre Connes had designed a variation on the Fabry–Perot that used a pair of curved mirrors. He showed that if the mirrors were spherically curved, and spaced so they focused light onto each other's surfaces, alignment should be a snap. Curved mirrors were easier to make than corner cubes or prisms, but Connes' plan required very precise spacing—within 0.01 millimeter—between the two mirrors. Jacobs and Rabinowitz didn't think that was practical, and abandoned the idea.

The military kept an interested eye on the TRG project. The Air Force dubbed it "Project Defender," reflecting hopes that the laser could fill a gap that terrified military planners—the lack of a defense against the Soviet Union's nuclear-armed intercontinental ballistic missiles. Cold War tensions and Soviet missile and space progress had created a balance of nuclear terror called mutually assured destruction. Military planners hoped that laser beams could zap Soviet ICBMs at the speed of light, 200,000 times faster than the comparatively leisurely speed of a bullet. It might be a long shot, but it was at least a hopeful one.

Military visitors streamed to TRG, looking for a future superweapon. Naturally they wanted to talk to the resident visionary who had coined the idea. But as long as he lacked a clearance, Gould couldn't tell them about TRG's latest results. It was hard to escape the irony. The research arm of the Pentagon considered Gould's idea vital to national security, but the security side of the Pentagon considered Gould himself a threat. Military scientists wanted him on the job, but security officers wouldn't let him in the door. Pentagon higher-ups couldn't or wouldn't take action to break the logjam and get Gould his clearance.

Throughout the last half of 1959 and into 1960, TRG thought the resolution of Gould's security problems was just around the corner. "We were expecting momentarily every week to hear Gould has got his industrial clearance. It was imminent and imminent and it never came," recalls Steve Jacobs. The closest moment came in the fall of 1959, when the New York office of the Industrial Personnel Security Review Board approved Gould's clearance. Yet that wasn't enough. Final approval had to come from headquarters in Washington, which insisted that Gould and his lawyers come to a final interview, which the security board kept postponing.

Yarmolinsky and Leventhal kept coaching Gould, warning him of "touchstones and shibboleths," a catalog of sensitive political issues considered litmus tests of loyalty. Admitting that he favored unilateral disarmament was tantamount to carrying a current Communist Party membership card. The two urged him to read J. Edgar Hoover's *Masters of Deceit*. Gould listened, but he was too stubbornly independent to stop questioning the establishment openly. He tended to be vague on details that might have cleared his reputation. A dozen years later, he would have been viewed as a free spirit with some rough edges, but in the Eisenhower era, he was a dangerous nonconformist.

Gould had loyal friends, but he also had an unfortunate knack for making powerful enemies, particularly Charles Townes, who had strong connections in the military research community. The central problem was that neither man really trusted the other, and both were after the same prize. To the intensely focused Townes, Gould seemed lazy, undisciplined, and irresponsible; a mediocre student trying to take credit for other people's ideas when he couldn't settle down to finish his own dissertation. To the free-spirited Gould, Townes was a pillar of the establishment that he had learned the hard way not to trust. Each man knew that he had conceived of the laser idea, so it was natural for each to suspect the other had stolen it. To further aggravate the situation, both Townes and Gould tended to be cautious and protective of their own interests around people they didn't trust. Townes was deeply principled, not the sort of man likely to sabotage Gould's clearance to try to block him from making a laser. Yet Townes might have considered it a matter of principle to warn security officials if he considered Gould to be untrustworthy.

Some of Gould's allies from TRG suspect Daly may have done something to hurt the clearance application. Security officials must have talked with Daly, but it's impossible to know what he said to them without access to the records. Daly was politically conservative, but his politics didn't affect his dealings with people. His hard-driving management style did get in the way, and it clearly irritated others at TRG besides Gould. Yet Daly's son recalls that his father liked Gould and credited him with inventing the laser.

Others may have shared reservations about Gould. The security clearance procedure was murky, and could be tainted by politics or false accusations. Under J. Edgar Hoover, the FBI collected files on a wide range of people accused of being subversive, including civil rights advocates. Gould had been on their suspect list long before the laser; it later came out that the FBI had tapped Gould's phone in the 1940s.

The security bureaucracy found itself firmly impaled on the horns of a dilemma. Gould's ideas had great military importance. Yet Gould's record was suspect. Security officials are paid to be paranoid, and military officials are taught to follow the rules. The rules said that Gould was a security risk who should not be cleared. The bureaucracy did what bureaucracies do best when faced with difficult decisions—it stalled, repeatedly postponing the Washington hearing through the winter.

Gould stewed in his isolated office as the delays dragged on, and petty annoyances grew aggravating. The layout at Two Aerial Road meant the path from his office to the men's room went though the secured area. Nobody worried too much as long as it seemed temporary, but as time passed something had to be done. TRG called in a contractor to tear down a wall so Gould could use the toilet without violating security.

12

THE SIREN CALL OF THE LASER

THE SIREN CALL OF THE LASER continued to spread through the little world of physics in the latter part of 1959. Scientists instinctively sniff the intellectual winds for hot new ideas in their fields. Ted Maiman had picked up on the laser scent quickly because he had just completed his last project successfully. Trying to build a laser was a tempting idea for those wanting a change, because they were finishing old projects, starting new jobs, or simply getting bored or frustrated with old lines of research that had yielded few results. Many talked about the idea. A few, like Nicolaas Bloembergen, looked closely and resisted the temptation because they knew they couldn't win the race. And some big industrial labs began venturing into the laser race in a serious way.

One was IBM, already the giant of the young computer industry. IBM had its roots in business machines that tabulated information on punched cards, but had jumped quickly into computing after World War II. In 1945, IBM established its first laboratory to study computing near Columbia University in Manhattan. As the computer industry grew, and solid-state electronics began to replace vacuum tubes, IBM broadened its scope to physical research, and opened a new research laboratory in Poughkeepsie, New York, about 60 miles up the Hudson River from Manhattan. One of IBM's projects was developing microwave spectroscopy as a tool for studying solids, pursued by a small group in Poughkeepsie headed by William V. Smith.

The laser idea had caught Smith's eye before the Shawanga Lodge meeting, held an easy drive from Poughkeepsie. One young physicist on his staff, Peter Sorokin, was invited to talk about his research on microwave masers. A second young physicist, Mirek Stevenson, went as well. The idea of making a laser captivated them, and when they came back from Shawanga Lodge, the two put decided to their other work on the shelf and turn to the laser. Smith encouraged them to go ahead.

Sorokin and Stevenson's background almost demanded they try to make a laser. Sorokin had earned his doctorate under Bloembergen at Harvard. Bloembergen was a rigorous teacher, but he rewarded excellence and hard work, and his students were well regarded. Stevenson had studied under Townes at Columbia. The pair complemented each other. Sorokin, the son of an eminent Harvard sociologist, was the idea man. Gentle and reserved, he sometimes had an otherworldly manner, but always had a sharp mind. A Czech immigrant originally named Cevcik, Stevenson was the assertive and practical one, who figured out how to implement Sorokin's ideas and made things happen. When he Americanized his name, he borrowed that of Democratic presidential candidate Adlai Stevenson, whom he greatly admired. With a head for business, Stevenson was determined to become a millionaire by age 35, and started a mutual fund on the side while working at IBM.

Sorokin quickly assimilated what he felt was the central message of the Schawlow-Townes paper—a formula that showed how fast atoms had to be excited for the laser to work. That led him to a further insight. He realized that the energy needed to excite a laser increases as the time light spends bouncing back and forth inside the laser cavity decreases. The time a photon spends in the laser cavity depends on how much light the mirrors reflect. The higher the reflectivity, the more times an average photon can bounce back and forth before it finally escapes from the cavity in the beam. If only one percent of the light is transmitted out in the beam, that meant the laser needed to amplify the light only a little more than one percent as it bounced back and forth between the mirrors.

That was a vital insight. Physicists expected lasers to amplify light only weakly on each pass, so in order to oscillate, a laser would need highly reflective mirrors. Sorokin and Stevenson then asked optics specialists how reflective mirrors could be. The answer they got wasn't encouraging. The specialists they talked with weren't sure a mirror could reflect more than 90 percent of the incident light. That assessment was overly pessimistic. A standard textbook of that era said clean silver and aluminum surfaces reflect more than 90 percent of the visible light striking them. The state of the art in the laboratory was considerably better; special coatings had been developed that could make surfaces reflect close to 99 percent of incident light, although the coatings were not very durable. Yet not knowing those possibilities, Sorokin came up with a way to get more reflective surfaces in a solid-state laser.

Sorokin's idea was to take advantage of a trick of optics called total internal reflection. The phenomenon is actually the flip side of refraction, the bending of light rays as they go between different transparent materials. The degree of bending depends on a quantity called the refractive index, which measures the ratio of the speed of light in a vacuum to the speed of light in the material. The speed of light in a material is always slower than in a vacuum, so the refractive index is always larger than one. Light can always go from a material with low refractive index, like air, into one with higher index, like glass, but it can't always go the other way. If light inside glass hits the surface at too steep an angle, it can't escape into the air, and is totally reflected back into the glass. The reflection is total, with no loss at all. The easiest place to see it is in diamond, which has a higher refractive index than glass, but you can see it in pieces of cut glass or crystal as well.

The attraction of total internal reflection was that it's total. All the light that hits the surface at a suitable angle is reflected back into the crystal because the light can't enter the air at that angle. It was the perfect mirror that Sorokin sought. He designed a square crystal resonator with faces at right angles to each other. Arrange it properly, and light hitting one face at a 45-degree angle would be reflected back into the crystal at 45 degrees, then hit the next face at the same 45-degree angle. Then the light would keep on bouncing around the resonator. Sorokin figured he could cut a piece off one corner of the polished crystal so some of the light could emerge.

Smith noticed something else when Sorokin described the idea. If the crystal had a refractive index just a little bit above $1.414 \ldots$, the square root of 2, the light would oscillate inside the crystal in a single simple pattern called a mode. That was an important advantage because it promised more control over the light generated by a laser. At that point, there were uneasy worries that the light, which bounced back and forth between a pair of mirrors, would dissipate into a multitude of modes instead of form a single coherent beam, so a single-mode cavity looked very promising.

Sorokin soon identified a suitable crystal, calcium fluoride or fluorite, which had a refractive index about 1.43. It happened to be the same crystal host Garrett and Kaiser were investigating at Bell. Like pure glass, calcium fluoride alone wouldn't make a laser because it didn't contain atoms that could be excited and then stimulated to emit light. It would have to be doped with another element that could be excited, as chromium could be excited in pink ruby. At Shawanga Lodge, Schawlow had recommended finding a crystal in which the lower laser level was above the ground state, a four-level laser. Sorokin and Stevenson took his advice, and started looking for suitable atoms they could add to calcium fluoride crystals. Ideally, the atoms should absorb light at a wide range of wavelengths, so a lamp or other source of ordinary light could excite them.

To find suitable materials, Sorokin and Stevenson dug through archives of spectroscopic research. Fortunately, Russian scientist P. P. Feofilov and his students had extensively researched light emission and absorption in elements that had been added to calcium fluoride. Two looked particularly promising for four-level lasers. One was uranium, which emitted infrared light at a wavelength of 2.5 micrometers. The other was samarium, which emitted light at 0.708 micrometers, a wavelength often classed as infrared, but faintly visible to the human eye. Both could be excited by visible light from readily available lamps. Both required cryogenic cooling, because the thermal energy at room temperature was enough to raise the added atoms to the lower laser level, blocking a population inversion. However, that didn't bother Sorokin and Stevenson, who weren't as concerned as Maiman about building a practical laser.

The drawback in searching for ideal crystals was that they weren't available off the shelf or in laboratory drawers, as ruby crystals were at Bell Labs. The calcium fluoride crystals had to be custom-grown, with uranium and samarium added in the right concentrations and the proper chemical state. Growing optical-quality crystals is always tricky, because the material must be made smooth and uniform, without flaws that could bend or scatter light, as objects floating on the surface of a still pond can deflect ripples. It was mid-May 1960 before Sorokin and Stevenson finally received their crystals. Then they had to send them out to another company to be cut and polished into the right shape.

The laser idea also had spread to the more traditional world of optics. American Optical, a large century-old manufacturer of spectacles and microscopes in

Southbridge, Massachusetts, had started a research lab in the early 1950s in an effort to develop new technology. Its first big project was a new system to film and project wide-screen motion pictures, but the company also sought hot new research areas, including the then-new field of imaging through bundles of optical fibers. Seeking to expand its scope in early 1959, the company hired Elias Snitzer, a former professor sacked the previous year by the Lowell Institute of Technology for refusing to cooperate with an investigation by the House Un-American Activities Committee. His leftist background didn't particularly bother American Optical managers. Like Bell Labs scientists, Snitzer didn't need a security clearance because the company was funding its own research.

Snitzer brought a fresh view to American Optical, instantly recognizing that optical fibers were guiding light in the same way that metal tubes called waveguides guide microwaves. He then did a more thorough analysis that explained important details of how fibers guided light, and American Optical management encouraged Snitzer to give a series of talks on his work. By then the idea of the laser was in the air, and it inevitably came up in questions after his talks. When asked what optical waveguides could be used for at a meeting of the New England Section of the Optical Society of America, Snitzer suggested lasers. When Snitzer talked at Columbia, Townes suggested putting light-emitting atoms only into the central core of the fiber, the region which guided light.

Since he wasn't working on microwave masers, Snitzer was not invited to the Shawanga Lodge conference, but he soon was caught up in the optical laser excitement. Working at an optics company, his logical starting point was glass. Optically perfect glass was easier to make than flawless crystals, and American Optical could make its own samples, reducing the time that Snitzer had to wait for samples to be prepared. Snitzer and two colleagues began searching for elements that would emit visible light when small quantities were added to glass.

American Optical decided to try making a visible-light laser because it would be easier to observe with photographic film or optical instruments than invisible infrared light. Snitzer's group started by studying the colored glasses made for use as optical filters, which get their color from elements added to normally clear glass. They bought samples of all the filter glasses made by the two largest manufacturers, Corning and Schott. They also tried some plastics. Their measurements identified four particularly promising elements—dysprosium, europium, samarium, and terbium—all members of the group of chemically similar elements called rare earths. They made up one-pound batches of glass containing each of the elements, and carefully measured their properties.

The attraction of rare earths was the structure of their energy-level ladders, which made them appear good prospects for producing a population inversion and stimulated emission. Rare earth atoms absorb light over a broad range of wavelengths, not just at a single band. This makes them a good match to ordinary lamps, which emit the range of wavelengths we see as white light. Many of these rare earth atoms linger a while in an excited state, helping to produce a population inversion. TRG and Bell Labs were experimenting with rare earths. Maiman had studied gadolinium, a rare earth. The samarium and uranium atoms that Sorokin and Stevenson were studying behaved like rare earth atoms when dispersed in crystals.

After testing all four rare earths, American Optical decided europium was the most promising laser candidate. They pondered how to determine if their samples were emitting laser light rather than ordinary fluorescence. Taking advantage of their experience with fiber optics, they stretched their glass samples into fibers, which they wrapped around lamps so the light would excite the rare earth atoms. Yet they were feeling their way in a whole new field, and their progress was slow.

So was the progress of other groups who had begun looking seriously at lasers. Few universities other than Columbia made a serious effort; like Bloembergen, most thought the task of making a laser was beyond the resources of a university. The most serious interest came from the high-technology companies of the time, particularly the aerospace and electronics firms like Hughes and IBM, which had already been working on microwave masers. Researchers at Raytheon in the Boston area looked at ruby. There was serious interest at the United Aircraft Research Center in East Hartford. Scientists from a small army of other companies also had listened to the new ideas at Shawanga Lodge: General Electric, the Martin Corporation (now part of Lockheed-Martin), Philco, RCA Laboratories, Sperry Gyroscope, Texas Instruments, Sylvania Research Laboratories, and Varian Associates.

A few researchers began exploring exotic-sounding ideas. One was the possibility of making lasers from semiconductors. The transistor was a decade old, but little research had been done on light emission from semiconductors, and efficient light-emitting diodes (LEDs) had yet to be invented. French physicist Pierre

Aigrain had suggested that passing an electric current through a two-terminal semiconductor device called a diode might generate laser light in 1958, but he thought of using germanium, which would not have worked. Benjamin Lax and Herb Zeiger, then at the MIT Lincoln Laboratory, realized other semiconductor materials might work. Nikolai Basov in the Soviet Union also began exploring the possibility of semiconductor lasers. But the ideas were far from practical.

Others like Irwin Wieder put the laser at the back of their mind as they worked on other projects. His focus at Westinghouse remained optical pumping of microwave masers. He didn't worry much about the low fluorescence efficiency he had estimated for ruby, because it wasn't a critical problem on the path to his immediate objective, demonstrating optical pumping of a maser. The ruby fluorescence source he described in the November 1959 issue of the *Review of Scientific Instruments* was a step toward his goal. As long as he got enough light for his optical pumping experiment, the efficiency of the ruby light source didn't matter.

The next step was to show that the red fluorescence could excite chromium atoms to the energy levels involved in powering a microwave maser. To do that, Wieder illuminated a ruby crystal with both red light and microwaves, and watched for changes in the absorption of the red light when he switched the microwaves on and off. He finally observed that effect on October 7, 1959. It was the first time anyone had demonstrated optical pumping in a solid, a significant research milestone. Wieder's report was deemed important enough for publication in the highly selective *Physical Review Letters*. He eventually received a patent on the idea. However, his progress stalled when he tried to take the further step of using optical pumping to power a microwave maser. He kept trying through the rest of 1959 and most of January 1960, but then slacked off.

Life had intervened. Westinghouse was based in Pittsburgh, and it was still a grimy industrial city where soot from steel mills blackened the snow. Pittsburgh winters could be bleak, and Wieder and his wife had grown tired of the place. They had both grown up in Cleveland, but had found the San Francisco Bay area a more pleasant place during Wieder's days in graduate school at Stanford. They decided to move back west, so Wieder applied for jobs in the bay area. Physicists were in demand, and he landed several job offers. He picked Varian Associations, because the company promised to help him set up a lab equivalent to the one he had at Westinghouse. He hoped to get restarted quickly in the new environment.

13 THE CRITICAL QUESTION OF EFFICIENCY

THE NOVEMBER 1959 ISSUE of *The Review of Scientific Instruments* carried Wieder's description of his ruby light source, where he wrote that the output he observed "corresponded to photon efficiency of 1%." Ted Maiman thought that low energy-transfer efficiency in ruby was the biggest roadblock to making a laser. The chromium atoms in ruby strongly absorbed visible light in two broad bands, blue and violet light centered at about 0.42 micrometers, and green and yellow light centered near 0.55 micrometer. Yet Wieder's paper said that only about 1 percent of the photons that a ruby crystal absorbed at those wavelengths excited chromium atoms to emit red light.

Initially, Maiman didn't doubt Wieder's published value. Others cited similar numbers. When Schawlow proclaimed that ruby wouldn't work in an optical laser, he cited experiments at Bell Labs that measured efficiency of one to ten percent, consistent with Wieder's results. Both sources seemed credible, and using their numbers, Maiman's calculations agreed with Schawlow's conclusion that ruby wouldn't work.

What bothered Maiman was not the numbers, but what happened to the missing energy. He, Asawa, and D'Haenens had made a series of spectroscopic measurements and a detailed analysis of the results to try to find where the energy went. When those experiments didn't pin down the mystery of the missing energy, Maiman decided to try measuring the efficiency in his own laboratory, to see if that provided any clues. Once he found where the energy was going, he hoped he could find a better solid-state material, since he was convinced that solids would make the most practical lasers.

His particular target was what physicists call the quantum efficiency. It measures what fraction of the photons absorbed by the ruby result in the emission of a photon of red light on the chromium emission line. If the process was perfectly efficient, and every absorbed photon excited a chromium atom that in turn emitted a red photon, the quantum efficiency would be 100 percent. That doesn't mean all the energy comes back out, because the photons emitted as fluorescence inevitably have less energy than the absorbed photons, but Maiman could account for that loss of energy.

Maiman did not try to repeat Wieder's exact procedure. Maiman liked to do things his own way, and he wanted to measure efficiency more precisely. In addition, using a different procedure would give an independent check on Wieder's measurement technique, a standard way of resolving scientific discrepancies.

The first step was to measure the absorption and emission of light by the chromium atoms in ruby. To do that, Maiman tuned his monochromator to emit yellow light at 0.56 micrometer, a wavelength strongly absorbed by chromium. He aimed that light through a quartz rod onto a one-centimeter cube of pink ruby. Some yellow light went straight through the little ruby crystal, and he could measure that fraction directly. His real concern was what was happening to the light energy left behind in the ruby crystal. Sapphire is essentially transparent to yellow light, so he knew it must be absorbed by the chromium atoms. But where was it going after the chromium atoms absorbed it?

One possibility was that the chromium atoms were releasing all the energy as a single yellow photon of exactly the same wavelength as the absorbed light, and dropping back to the ground state in which they had absorbed the light. That's a process called scattering, and it's the same thing that happens to light trying to pass through a cloud; the light emerges in a different direction than it entered. A second possibility was that the excited chromium atom released a little bit of the energy into the crystal and dropped to the slightly lower energy level, which emitted the red fluorescence that Maiman hoped to use in a laser. A third possibility was that all or most of the energy leaked away from the chromium atom to heat up the crystal without ever being released as light.

To see what was happening, Maiman and D'Haenens set light detectors on faces of the ruby crystal at right angles to the path of the yellow light. One looked for light scattered at the yellow wavelength; the other looked for the red fluorescence. By comparing the light intensities, they could measure how quickly the chromium atoms released energy in different ways, and how efficient the process was.

The results were a pleasant surprise. They found little of the yellow light was scattered from the crystal. Even less energy was leaking into the crystal as heat without being turned into light. Instead most of the excited chromium atoms were dropping into the lower-energy state that emitted red fluorescence, then emitting red photons. Maiman's first rough estimate of the quantum efficiency was 70 percent, and that was very good news. Later more precise measurements showed that the quantum efficiency actually was close to 100 percent. Ruby fluorescence was far, far more efficient than either Wieder or Schawlow had reported.

The new experiments convinced Maiman that he understood energy transfer within ruby. His measurements of high quantum efficiency put ruby back on the list of viable candidates for either an optically pumped microwave maser on an optical laser. Maiman looked at the stakes, and decided to take the bigger gamble, and go directly for the ruby laser.

It was not a decision that George Birnbaum or Harold Lyons embraced enthusiastically. The two managers reminded Maiman of Wieder's discouragingly low efficiency for ruby, and Schawlow's doubts about ruby. Birnbaum says they didn't tell him Maiman to stop working on ruby, and that he and Lyons strongly supported the laser project. But their continuing doubts that ruby was the best choice for a laser material fueled a growing friction that was particularly intense between Maiman and Lyons. It was natural to say nothing that might alert potential competitors to the possibility that ruby might work. Yet Maiman also kept his own managers in the dark as he began serious work on the ruby laser.

His action highlighted the growing dysfunction of Hughes management. The company had hired the best and brightest scientists they could find, and they often came with strong or high-strung personalities. Andy Haeff, a top engineer who was director of Hughes Research Laboratories, was on his way to a nervous breakdown. Maiman usually got along well with small groups of his peers and people who worked for him, but his relationships with management were uneasy. He had a tendency to be headstrong and had the courage of his convictions to question management decisions. Even worse, Maiman had a tendency to be right. When he decided to charge ahead with the ruby laser, he also decided the easiest way to avoid clashes with Birnbaum and Lyons was to avoid telling them what he was doing.

A serious effort to develop a ruby laser required a suitable light source to excite the chromium atoms. The requirements were demanding because ruby is a three-level laser system, with the ground state also serving as the lower laser level. To produce the population inversion needed for stimulated emission to dominate inside the laser cavity, more than half of the chromium atoms had to be excited out of the ground state, so more atoms were in the upper laser level than in the ground state. That meant the light source had to be extremely intense. While Maiman worked on an analytical model of how the laser would operate, he asked Asawa to look for a suitable light source.

It was clear any light source had to be extremely bright, with an extremely high power per unit area, if it was to excite more than half of the chromium atoms. Ideally, it should excite almost all of them to the upper laser state. It wouldn't do to focus light from a large source onto a small area; the intensity had to be extremely high at the surface of the lamp itself. It also had to generate a steady light, because like virtually everyone else, Maiman initially envisioned a laser that would emit a continuous beam.

One possibility was passing a very strong current through an air gap between a pair of carbon electrodes, called a carbon arc. The carbon arc was the first bright electric light, and had been around for more than a century. But it also had serious drawbacks that made it impractical for most applications. It produced noxious fumes, and Asawa decided it was too hot for use in a laser. He also considered an electric arc passing between a pair of zirconium electrodes, but zirconium arcs were mounted inside large bulbs to keep them from sputtering zirconium over everything, and the bulb kept the arc too far away to produce the intensity needed on the ruby crystal.

Movie-projector lamps were another possibility. They have to be very bright because the projector spreads the image of a small piece of film onto a giant screen in the theatre. They are so bright that if a movie-theatre projector stops, the heat from the bulk quickly melts the film. Standard projector bulbs wouldn't do because they don't produce enough light at the blue-violet and green wavelengths needed to excite chromium atoms. However, from his work at UCLA, Asawa knew about a special compact lamp still manufactured today for industrial use. Called the AH6, it draws 1000 watts of electricity and excites mercury vapor to emit blindingly bright blue-violet and green light. Much of that energy goes into heat, so it self-destructs unless it is cooled by flowing water around it, or by blowing air past it at high pressure. Even with the required cooling, the bulbs burn out in only about 25 hours. But it was the most intense bulb Asawa could find, and he recommended it to Maiman.

The need for cooling and the lamp's short lifetime were not fatal flaws. Maiman's immediate goal was to demonstrate a working laser, not to build something ready for everyday practical use. In fact, most other laser schemes were considerably more complex, and required custom-built equipment to excite the laser material. The AH6 looked like a reasonable starting point. If it worked, Maiman figured he could look for simpler approaches later.

As an alternative, Asawa also looked at light sources which generated very bright flashes. The attraction of a pulsed source is that for a brief instant it can generate far more light than any steady source can without burning out. After searching through references, he began a series of experiments in which he passed strong electrical pulses through wires, making them explode in bright flashes of light. The flash ran the whole length of the wire, so he could make a long, thin light source to illuminate the entire length of a ruby laser rod. However, exploding wires were messy, and didn't look like an ideal source.

In November, Asawa described his search for pulsed light sources and the exploding-wire experiments to one of his office mates, Leo Levitt. It happened that Levitt, an amateur photographer, had recently bought a professional flash-lamp for his camera, and had brought his expensive new toy to work to show to his friends. After Asawa finished, Levitt pulled out his fancy flashlamp and asked, "What do you think of this, Charlie?"

Asawa was impressed. Most amateur cameras at the time used small flash-bulbs that could be fired only once, igniting metal wire that flared inside a bulb. The photographic flashlamp was elegant in comparison. An electronic circuit built up an electric charge, then fired a pulse through a sealed tube containing xenon gas. The circuit controlled the shape of the electrical pulses to produce the brightest possible flash from the xenon lamp. Both the circuit and the lamp could be fired again and again, limited only by how fast the circuit could charge. It was the sort of neat technology that physicists could appreciate. It also produced the brightest flashes of light that Asawa had ever seen.

Asawa showed the flashlamp to Maiman, who was also impressed by the lamp's intensity. However, it was not an ideal source, and Maiman still clung to the hope that he could find a bright source of steady light so he could excite a laser to emit a continuous beam. Other groups were taking that approach, and it seemed to promise a more useful type of laser.

Maiman continued to consider other materials, including the gadolinium crystals used in microwave masers. Gadolinium emitted strong ultraviolet fluorescence rather than the red fluorescence from ruby, and Maiman was briefly intrigued by the idea of making an ultraviolet laser. However, the ultraviolet gadolinium laser would have been a three-level laser system, like ruby, and his calculations of its energy requirements weren't encouraging. By November 1959, he was concentrating on ruby.

Ruby had its own problems, Maiman found when he sat down to analyze the prospects for pumping a ruby laser with the brightest available continuous light source, the AH6 lamp. A big one was how to deliver as much light as possible from the lamp to the laser rod.

The lamp was a cylinder, which emitted light from its sides in all directions. The best design for a ruby laser was a cylindrical rod with mirrors on both ends, and pump-light entering from the sides. The problem was transferring light between the two cylinders. Put the lamp beside the laser rod, and some light would enter the rod, but most would radiate away in other directions. To make a laser, Maiman would have to collect that light and focus it back onto the laser rod. Fortunately, the geometry of an ellipse offers a simple solution. Light that bounces outward from one focus of a reflective ellipse will bounce back through the other focus (fig. 13.1). Maiman saw that he could use this trick by lining the cylindrical lamp and the laser rod inside a reflective tube with an elliptical cross section. If the lamp is lined up along one focus of the ellipse, the elliptical tube will focus the light it emits onto the other focus. Put the laser rod at the other focus, and the pump-light will converge on it. Townes had suggested his students try the same geometry for exciting a potassium laser at Columbia.

The elliptical cavity is a simple and ingenious design that is still used to pump solid-state lasers today. It made the best possible use of the light from the intense AH6 lamp. Yet even with that advantage, Maiman calculated a ruby laser pumped by the AH6 would barely work. As a man who took pride in his creations, he wrote, "it was hard to get very excited about a marginal design."

Seeking new insight, Maiman turned back to his analytical model and tried a different way of calculating what sort of pump source he needed. This time he

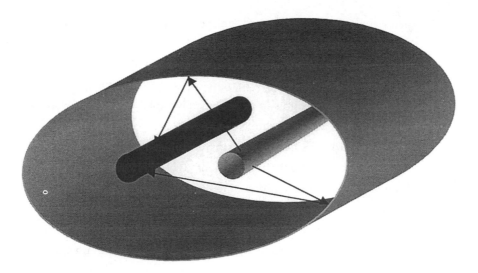

FIGURE 13.1. Elliptical pump cavity reflects pump light from a lamp (light cylinder at right) to the ruby rod (dark cylinder at left).

tried to estimate a quantity called the black-body temperature that a lamp should have to optically excite enough chromium atoms in the ruby.

The black-body temperature is a way to measure the wavelengths where a light source emits the most light. (A "black body" is a theoretically ideal object that absorbs and emits all wavelengths with 100 percent efficiency.) Every object above absolute zero radiates some energy, but the radiation may not be at wavelengths we can see. The hotter the object, the more energy it radiates and the shorter the peak wavelength. Our bodies radiate invisible infrared light, with a peak near a wavelength of ten micrometers, but no detectable visible light. A burner on an electric stove is hundreds of degrees hotter, so emits more light at shorter wavelengths, and glows cherry-red. The filament of an incandescent bulb is much hotter, about 2500°C, and glows white to the eye although it's brighter in red light. The sun also looks white, but it's much hotter, around 5800°C, so sunlight contains more of the shorter blue and green wavelengths.

Chromium atoms in ruby absorb violet, blue, and green light, so they need a light source with a high black-body temperature. Maiman calculated the requirement for a ruby laser was roughly 4700°C. That looked discouragingly high, until he realized that the black-body temperature of photographic flashlamps could reach about 7700°C. That made the xenon photographer's flashlamp looked very good.

Maiman went back to his model and confirmed that a pulsed lamp would produce enough light energy to excite a ruby laser. His model showed that the key parameter was brightness, the power per unit area delivered to the surface of the ruby crystal. The wavelengths also were right. Xenon flashlamps emitted strongly at green and blue wavelengths to properly expose color film. Those were just the bands needed to excite chromium atoms in ruby.

A laser excited with a flashlamp inevitably would be pulsed, because it could produce stimulated emission only in the instant after the lamp's brief flashes had inverted the population of chromium atoms. No one else in the laser race was seriously working on pulsed lasers. They generally envisioned optical lasers as types of oscillators, which produced steady signals at lower fixed frequencies, so they were trying to produce a steady optical oscillation. It was an important difference, but Maiman decided it wasn't a major problem. His first goal was to demonstrate laser operation, and pulses would be sufficient to show a laser could work. Besides, as with flashlamps, pulsed beams might serve different purposes than steady ones.

That choice was an important step because it took Maiman down a path different than Schawlow and Townes had envisioned for laser development. It wasn't immediately clear where the path would lead, but equipped with his fresh vision of a pulsed laser, Maiman charged ahead.

A search through industrial catalogs showed that the flashlamps producing the most intense light were three models made by General Electric that all looked like coiled springs. The mid-sized version, the FT-503/524, was similar to one that Harold Edgerton had developed at MIT for low-level aerial photography early in World War II. General Electric also offered the larger and brighter model FT-623, and a smaller version called the FT-506. The lamps were so bright they left a comfortable safety margin above the minimum requirement Maiman calculated that he needed. Even the smallest, the FT-506, promised up to three times the minimum intensity. He ordered several lamps of each size, but started to design his laser around the smallest lamp, which required much lower electrical power to drive it than the larger ones.

Although the numbers looked good, Maiman worried that he might be missing something. Doubts about the prospects for ruby had spread beyond Bell Labs. Both Birnbaum and Lyons had echoed Schawlow's warnings. Peter Franken, who was a friend of both Maiman and Gould, had largely concluded the laser wouldn't work, and was getting ready to go on the record with his doubts. To reassure himself, and perhaps to convince Birnbaum and Lyons, Maiman decided to perform a new set of measurements.

He wanted to see what happened to the ground state in ruby when it was illuminated by flashes of light from a flashlamp. He figured that he could see by monitoring how the ruby crystal responded to microwaves as the light flashed on and off. He discussed the idea with Asawa's advisor at UCLA, spectroscopist Robert Satten, who made his own calculations that verified Maiman's and helped Maiman refine his plan.

To measure what happened, Maiman and D'Haenens placed a one-centimeter cube of pink ruby between a pair of parallel metal plates. They spaced the plates so that microwaves at a frequency of 11.3 gigahertz would resonate within the cavity they formed, passing back and forth through the ruby crystal. Chromium atoms in the ground state absorb that frequency, so any changes in the ground-state population should show up as changes in the microwave intensity.

Wieder had looked at changes in microwave absorption in a ruby crystal illuminated with red light from another ruby, to see if the red light had excited chromium atoms in the crystal. Maiman was looking for something different, the effects of the bright violet-blue and green light from the flashlamp. In his experiment, a pair of bent quartz rods served as light pipes, collecting light from an FT-506 flashlamp. Total internal reflection trapped the light in the rods, and guided it to two other faces of the little ruby cube. Other optical equipment was set up on the third pair of faces to probe the ruby with certain wavelengths and measure what happened. The flash of pump-energy lasted just 0.2 thousandths of a second.

The test was a crucial one. If Schawlow was right, light from the flashlamp would have no discernible effect on the ruby crystal, and the microwave intensity would not change during the pulse. Yet Maiman saw a pulse rising sharply on his oscilloscope screen and falling gradually. The microwave intensity had peaked briefly while the flash illuminated the ruby, then dropped slowly. The sharp rise showed that about three percent of the chromium atoms had been removed from the ground state; the slow decline in microwave intensity came as the excited chromium atoms dropped back to the ground state. Maiman was delighted because that value was close to what he had predicted before the experiment.

The flashlamp pulse had excited only three percent of the chromium atoms out of the ground state, but that was far from the peak efficiency possible. The quartz light pipes collected only a small fraction of the light emitted by the flashlamp.

Some light leaked out along the way to the ruby crystal. Maiman believed that if he could deliver all the light from the flashlamp onto a ruby crystal, he should be able to excite more than half of the chromium atoms out of the ground state, creating the population inversion essential for laser action.

To nail down his case, Maiman repeated the experiment, but watched for changes in the absorption of light rather than microwaves. To do that, he used his monochromator to generate violet light at 410 nanometers, a wavelength strongly absorbed by the ground state of chromium in ruby, and shined that light into one side of the ruby cube. Then measured how much of the violet light passed through the ruby when the flashlamp fired. As in the microwave measurements, the amount of the violet light passing through the crystal increased for the 0.2 thousandths of a second that the flashlamp fired to illuminate the ruby crystal, then dropped back to its original level. The increase in violet transmission meant the ground-state population of chromium atoms had dropped by about three percent. That matched the results of the microwave measurements. Ruby was looking very good.

"The measurements were not especially difficult," recalls Viktor Evtuhov, a Hughes physicist who worked with Maiman. Yet pitfalls abounded because most of the Hughes team was not familiar with optical techniques. Other pitfalls had caught Irwin Wieder and Bell Labs. Maiman had eluded them by meticulous design of his experiments and careful execution of the measurements. Three separate experiments all pointed to the same result. Flashlamps could excite chromium atoms out of the ground state very efficiently. Ruby was a viable laser candidate. Maiman charged ahead.

The next step was designing a laser. As nearly as Maiman could measure, every chromium atom that absorbed a photon from the flashlamp pulse was excited out of the ground state—virtually 100 percent efficiency. That was unquestionably good news. But so far he had only been able to excite three percent of the chromium atoms out of the ground state because the photons from the flashlamp didn't reach them. He needed a way to deliver enough of the light in the flashlamp pulse onto the ruby crystal to excite all the chromium atoms. Maiman focused intensely on the problem, but other events got in the way.

Hughes Research had been located in the sprawling Culver City facilities of the parent Hughes Aircraft since the labs were established as a separate entity in 1953. With the rapid growth of the aerospace industry, the place had gotten crowded. Some groups were working in Quonset huts and converted aircraft hangers, and Hughes desperately needed more research space. In December 1959, the company leased a new building overlooking the Pacific Ocean in Malibu, a location that may be the most beautiful site occupied by any American industrial laboratory, and announced plans to move the labs there. As construction crews finished labs and offices in the new building, Hughes began moving research groups from Culver City.

The move was a big step up in ambience, but a nuisance for the little group trying to build a laser. When spring arrived, so did the word to pack. Laboratory equipment had to be disassembled, moved, then carefully reassembled. Offices had to be packed, moved, and unpacked. Work ground to a halt in the interim.

The move also complicated Maiman's personal life. Married and with an infant daughter, he had been in process of buying a house in Palos Verdes when the move was announced. Palos Verdes isn't a bad commute from Culver City, but it's much farther away from Malibu. Maiman had to cancel the deal and find a new house closer to Malibu.

Unable to work while his lab was being packed, moved, and reassembled, Maiman settled down to write a paper describing his measurements, which he submitted to *Physical Review Letters* in mid April. Titled "Optical and Microwave-optical Experiments in Ruby," it reported "the first observations of ground-state population changes in ruby due to optical excitation." On the surface, the change of only three percent seemed a minor triumph. Maiman did not mention his hopes that the measurements were a key step on the road to the laser. He was far too cautious for that.

Publishing the paper probably seemed a good way for Maiman to demonstrate to Hughes management that he was making progress. They knew that it was an important step toward his goal of a ruby laser. Others didn't. The editors of *Physical Review Letters* sent a copy to Schawlow for comment, part of the usual process of refereeing scholarly papers to assure they deserve publication. Schawlow gave the paper his blessing, but didn't understand all the implications at the time. Neither did the feisty editor of *Physical Review Letters*, Sam Goudsmit.

Maiman had picked *Physical Review Letters* for its quick turnaround, and the journal obliged by scheduling the paper for its June 1 issue. Maiman charged ahead designing a ruby laser, never stopping to think that someone like Goudsmit might not instantly recognize how dramatic of an advance a working laser would represent over a series of experiments with ruby.

14

AN IDEA SIMPLER IN THEORY
THAN IN PRACTICE

AS MAIMAN QUIETLY WORKED ON RUBY, optimism about prospects for making some kind of laser was spreading beyond the inner circle of researchers, managers, and project sponsors. The leading industry magazine *Electronics* listed the laser among hot new concepts in its March 11, 1960, issue. "Working infrared and optical masers for communications and radar-like purposes are expected within two years. A wide variety of solids, liquids and gases are being investigated." Yet the laser wasn't headline news. It was buried on the fifth of six pages of a preview of an Institute of Radio Engineers convention in a magazine written for electronic engineers. The prediction also was carefully qualified by adding the weasel words "are expected" to the optimistic projections that had emerged after the Shawanga Lodge meeting.

The editors of *Electronics* were seasoned professionals who understood the ways of research and development. They knew that all bright ideas didn't pan out. Indeed, the reality by March of 1960 was that laser progress at the big labs was slower than the optimists had hoped. Like life, laser physics was proving to be easier in theory than in practice. A number of skeptics besides Peter Franken were starting to have doubts about the prospects for building a laser. Some solid-looking ideas were not working out, or were dragging on indefinitely.

The first failure was firing an electric discharge into pure helium. John Sanders had hoped it might produce a laser when he came to spend his sabbatical at Bell Labs. He knew that some helium energy levels had looked as if they should be excited by electrons passing through the gas. On paper, the helium atoms should have remained excited long enough to produce the population inversion needed for stimulated emission and laser action. Yet his experiments had shown that they didn't. Only later did Sanders learn that other states of helium atoms absorbed light at the wavelength where he was trying to stimulate emission, an effect called radiation trapping. The discharge excited some helium atoms to those states, where they soaked up the stimulated emission as fast as the helium atoms in the upper laser level could produce it.

Sanders didn't give up when he returned to Oxford University. He put out word that he wanted a graduate student willing to tackle the laser problem. That caught the attention of Colin Webb, who had worked on a 10-centimeter microwave maser as an undergraduate in the summer of 1959. Webb hadn't heard of the Townes–Schawlow paper, but he realized that an optical laser might be possible, and jumped at the chance to work with Sanders, although he didn't start until October 1960.

Radiation trapping was only one of the pitfalls that awaited early laser developers. Other problems led both Columbia and TRG to abandon the potassium-vapor laser before Schawlow and Townes received U.S. Patent 2,929,922 for their optical maser proposal in March 1960.

Potassium looked simple, because potassium lamps emitted light at a wavelength that potassium vapor absorbed. Yet the potassium lamps emitted only weak light, making the experiments difficult. Frustrated, Abella and Cummins borrowed Townes's station wagon to haul their potassium apparatus down to Bell to study it with Schawlow's expensive spectrometer. The instrument revealed the absorption lines they expected, but none of the fluorescence they were hoping for. Townes suggested they put one of their linear potassium lamps in a reflective elliptical cylinder with a tube containing potassium vapor, the same approach Maiman proposed for pumping a ruby rod with a linear flashlamp, but that was little help.

Potassium vapor caused problems. Glass absorbed the vapor, darkening the tubes that were supposed to contain the metal vapor. Abella and Cummins tried using tubes made of sapphire, which potassium didn't darken it in the same way, but the sapphire wouldn't bond properly to other glass in their apparatus. They tried distilling the potassium to purify it, but the metal vapor reacted violently and blew up their purification apparatus. Other things went wrong. Frustrated by a lengthy catalog of problems, the two students looked for alternatives. Abella found that mercury vapor emitted an ultraviolet wavelength that potassium vapor absorbed. But he was more encouraged when he discovered that a 388.8-nanometer ultraviolet line emitted by helium overlapped a wavelength absorbed by cesium. He and Cummins seized the chance to switch to cesium. Like potassium, cesium is a highly reactive alkali metal, so it still had to be handled with care. Yet they could hope that cesium wasn't quite as nasty.

Seeing his students' frustration, Townes arranged for Schawlow to spend the spring term at Columbia as a visiting professor. By the spring term, the Columbia lab had bought its own expensive spectroscope, and Schawlow taught the two students how to use it. When experiments showed that cesium shared many of potassium's problems, Abella asked Schawlow about ruby. The crystal was a benign and simple material compared to the corrosive metal vapors. Abella had a sample, which glowed with deep red light when he illuminated it with a mercury lamp. Yet Schawlow insisted there was no way to excite enough chromium atoms to make pink ruby work. He helped them with spectroscopy, but he had no new materials to offer. Oliver Heavens could help them with optics, but not with laser materials. Reluctantly, Abella and Cummins returned to cesium, and to the graduate-student nightmares of a doomed dissertation project. Heavens made the most of his American sabbatical, traveling, visiting other labs, and giving talks that projected an overly rosy picture of Columbia's laser progress.

As the only open laser program at a university, Abella and Cummins found themselves hosting visitors from all over, including Nikolai Basov from the Soviet Union, and physicists from aerospace companies United Aircraft and Raytheon. Through his friendship with Steve Jacobs, Cummins invited Gould to see their experiment. Gould, who probably enjoyed the chance to talk openly about lasers, gave the two some helpful suggestions, but most of the visitors pumped the students for information. The students, naive in the politics of research, didn't realize they were in a race, and sometimes said too much. Abella was very annoyed when he found that one visitor had published an idea he had picked up at the Columbia lab.

TRG couldn't talk openly, but behind closed doors they also were making only slow progress. Many of their scientists were learning a new field and the company's font of innovation, Gordon Gould, was climbing the walls in his isolated office, without access to the secured laboratory or to the classified results it produced. Townes, Schawlow, or Javan might have productively analyzed theoretical concepts with pencil and paper, but Gould's strength was his skill at testing ideas experimentally and using the results to refine his concepts. Stuck in his office, he smoked like a chimney, and as time passed and pressures built, the inspirations that powered his inventiveness tended to slip out of reach.

The pressure came from Gould's looming security hearing. Leventhal and Yarmolinsky spent hours coaching him how to behave in front of security officials, a process that reinforced his native cynicism that the system was inherently corrupt. Frustrated at his treatment, Gould saw less and less chance that he might win. Not until April 19, 1960, did Gould get his first hearing at the Pentagon. With the lawyers at his side, he recounted his past activities yet again for the uniformed security officials. When it was over, he and the lawyers had to collect yet more affidavits and documents and forward them to the slow-moving security bureaucracy. Trapped in bureaucratic limbo, Gould drifted out of touch with the classified laser project. He always had problems finishing projects and meeting deadlines. The pressure made the problems worse, and Gould's well of innovative ideas ran dry.

With Gould out of the loop, key decisions were routed through Daly. That sometimes slowed up the works. After Rabinowitz realized that potassium was unsuitable for a laser, and had made measurements that showed cesium would be better, Daly wouldn't authorize the change until he heard from Heavens that Columbia was doing the same thing. Cesium was still a difficult metal, but it was easier than potassium, and Rabinowitz and Jacobs at last began to tame the system. The TRG scientists got a few chuckles when visiting military officials waxed wildly optimistic. "The generals would come and ask when we were going to have it," recalls Jacobs. "They wanted ruby logs," able to fire laser cannon-shots.

Progress stalled on the gas-discharge laser. Teamed with three theoretical physicists, experimentalist Ben Senitzky felt bogged down in theory. The TRG group had made the strategic mistake of thinking they needed to measure gain before demonstrating oscillation, not realizing oscillation was easier to see.

Materials problems side-tracked the TRG solid-state laser program. Ron Martin thought that a four-level material would make a better laser than a three-level material like ruby, and tried to identify a promising one. Schawlow echoed Martin's thoughts when he came to visit TRG, insisting that ruby wouldn't work and recommending crystals containing small amounts of rare earth elements as the best four-level laser candidates. Yet neither Schawlow nor TRG was close to finding an ideal material. Schawlow walked away from the TRG meeting discouraged. Later he wrote, "everything we discussed presented formidable problems, and an operating laser did not seem close."

Schawlow was talking a lot about lasers in early 1960, but he wasn't making much progress toward making one. His visiting professorship at Columbia was draining his energy. He faced a grueling daily commute from his home near Bell Labs in New Jersey to Columbia's uptown Manhattan campus. He was teaching courses as well as advising Townes's students. At best, he could manage a half-day a week at Bell. It was too much for Schawlow, who lacked Townes's drive and physical stamina. Schawlow found his time at Columbia "an exhausting ordeal," and later said it ended with him "quite ill with a succession of colds and an infection which took weeks to clear away."

Resting at home was problematic. Schawlow had three children, born in 1956, 1957, and 1959. His wife Aurelia stayed home with them, but preschool children drain all available parental time and energy. The couple also had another concern. The oldest child, their only son Artie, was not developing normally and didn't talk. He would eventually be diagnosed as autistic. Schawlow didn't talk about it at work; it wasn't the sort of thing people talked about in 1960.

In addition, Schawlow's heart simply wasn't in building a device. As a high school student during the Depression, engineering had looked like a good job, but in college and graduate school his interests had shifted to physics, where the goal is understanding nature rather than building things. By the time he reached Bell Labs, he was worried he might be assigned to an engineering job.

It all added up to Schawlow being far from peak form when *Physical Review Letters* asked him to referee the paper on ruby fluorescence that Maiman submitted in late April. Schawlow read the paper and urged the journal to publish it. Schawlow noticed that Maiman had succeeded in exciting a significant fraction of the chromium atoms in ruby, but he didn't think much about the implications, which Maiman glossed over. Schawlow guessed that Maiman might be trying to make an optically pumped microwave maser, but he was too tired and too busy to ponder on the matter. He also may have missed the implications because Maiman's measurements were made with brief, intense pulses of light. AT&T was in the communications business, so everyone at Bell was working on continuous-beam lasers suitable for communications. They sought a laser that, like a radio oscillator, would generate a steady beam at a fixed wavelength or frequency, which could be modulated to transmit a signal for communications. Pulses wouldn't fill the bill.

Back at Bell Labs, the helium-neon laser had emerged as the leader in the laser race. Savvy research managers like Al Clogston expected it would work, but it was slow going. Javan and Bennett worked long hours in the lab trying to nail down the details. After analyzing the spectroscopy and energy levels in helium and neon in exhaustive detail, they moved on to the critical question of whether or not the light gained power as it passed through the excited mixture of helium and neon. If it did, they were getting the stimulated emission they needed to move on to the next step of adding a resonant optical cavity to try to make a laser oscillator. If they saw no gain, they had a problem.

It was a properly methodical approach to science, taking one step at a time and making sure that it worked before moving on to the next. Gain came before oscillation, so Bennett and Javan thought they should demonstrate the first before they moved on to trying the second. But it put them in the same trap as Ben Senitzky at TRG, because their mixture of helium and neon could only produce low gain, so the increase in power would be small and very hard to measure. Going directly for oscillation would have been easier, because the power builds up to much higher levels in an oscillator. But that became obvious only with 20-20 hindsight.

They started with a light source that emitted near the wavelengths they expected their mixture of helium and neon to amplify, and aimed the light down a tube containing a mixture of helium and neon. They expected that passing an electric current through the gas should excite the helium atoms, which would transfer their energy to neon, producing a population inversion that would amplify the light. Comparing the light emerging from the tube with and without the electric discharge should indicate the gain, but they worried that the low gain was likely to be lost in the inevitable noise.

Javan came up with an ingenious-seeming plan to overcome that problem. He would modulate the intensity of the light source at one frequency, and the strength of the discharge at a second. In theory, the signal they were looking for should have a frequency that was the sum and/or the difference of the two modulation frequencies. That frequency should be easy to identify, and his idea took advantage of the fact that electronic measurements are often easier if the signal varies regularly in time than if the signal remains at a steady level. Yet the idea proved difficult to implement. Despite accumulating banks of expensive equipment, they struggled for months to collect reliable data. In early January 1960 both Javan and Bennett thought they saw gain, but they couldn't agree how much. Long and careful measurements did not yield consistent results, and they couldn't calibrate the apparatus reliably. Progress ground to a virtual halt.

Bell was very interested in the laser, but Javan had been working on the idea for over a year, and upper management was starting to get impatient. Bell had deep pockets, Javan had come highly recommended by Charles Townes, and the laser project was a high priority. But Javan's lab had become a money pit full of very expensive equipment that was producing very little obvious progress. As Schawlow recalled, "The research management at Bell Labs became concerned whether this was all a waste of a rather considerable amount of money or whether there was indeed some hope in it."

Each purchase had seemed a good idea at the time. When Javan arrived at Bell, his managers urged him to buy a large and expensive precision magnet for research on solids. Javan duly ordered one, but never did any solid-state research. He ordered more expensive equipment for his research on the helium-neon laser. As the bills mounted, top managers began to see Javan as a spendthrift, with an unfortunate tendency to change his plans after he ordered new equipment but before it was installed. The costly unused magnet collecting dust became a sore point. When Javan wanted more equipment during a budget freeze, he had to appeal all the way to the laboratory director, Hendrik Bode. Bennett later estimated the lab may have accumulated a million dollars worth of equipment, a huge sum at the time.

The expense might not have bothered managers as much if the equipment had been busy producing results. Physics was becoming a big industry, and Bell had money to spend on promising research. Yet the sight of expensive instruments sitting unused touched tender nerves in a generation of managers that had lived through the lean years of the Great Depression. Perhaps a culture gap yawned between them and Javan's upbringing in the Iranian upper class. When upper management began asking hard questions, Al Clogston, the manager in charge of Bell's laser research, insisted Javan and Bennett were doing good work. That was enough to satisfy the bosses for a while. Fortunately Bennett made an important step forward, which eased the pressure from above.

When Javan headed south to vacation in the Virgin Islands, Bennett decided to try a new way of measuring gain in the gas mixture. He set up a partly transparent mirror that split the input light into two beams, one going directly to a sensor, the other passing through a mixture of helium and neon on its way to a second sensor. Then he hooked both sensors up to an electronic circuit. Before applying any voltage to the gas mixture, he balanced the circuit so it gave zero output. Then he turned on a circuit that applied a discharge across the helium–neon mixture. The circuit varied the voltage applied across the helium–neon mixture at a particular frequency, so the light intensity should vary at that frequency if the discharge affected the light passing through the gas. Absorption caused the signal to drop; gain caused it to increase.

It's hard to spot such variations if they cause only slight changes in the intensity of light reaching a sensor. Bennett's circuit made them easier to spot by combining the output of the two sensors so the total added to zero when there was no discharge passing through the gas. Once he balanced that "bridge" circuit, any change in the signal moved it away from zero—making it much easier to observe. It's a classic way to measure a very weak signal, long used in electronics, and it worked for Bennett.

When Javan returned, the two tried it together, and for the first time saw a clear gain signal on the evening of March 31. Stimulated emission amplified the light in their tube by only 0.4 percent at a wavelength of 1.118 micrometers, they calculated. That wasn't enough to make a laser work. The best mirrors then available absorbed more light than the stimulated emission produced. But it proved that they had stimulated emission on a line they had predicted, making it a real milestone. It was also eased the fears of worried Bell managers.

Fox and Li began to get usable results from their computer simulation of light resonance in a laser cavity about the same time Bennett and Javan conclusively measured gain. Their initial results showed that a reasonable amount of light would bounce back and forth between a pair of small and perfectly flat mirrors placed at the opposite ends of a laser tube. The flat surfaces of the mirrors had to be exactly parallel to each other, and precisely perpendicular to the hollow axis of the tube. The light had to be along the axis of the tube, or it would eventually bounce out the side. The mirror reflectivity had to be high. The conditions were difficult to achieve, but they weren't impossible.

Critically, Fox and Li's calculations also alleviated a major worry that had led some to doubt prospects for the laser. A natural effect called diffraction scatters light at sharp edges, such as the sides of the mirrors or the tube. Some researchers had expected that diffraction would scatter so much of the light bouncing between a pair of small mirrors that laser oscillation would be impossible., Instead, Fox and Li showed that the light intensity would peak at the center of the beam emerging from the partly transparent mirror, dropping toward the edges.

The progress was encouraging, but much more work remained to build a laser. Javan and Bennett needed to refine their gain measurements and increase the strength of the stimulated emission. They could have tried to publish a paper reporting gain, but decided instead to wait until they had a working laser. Fox and Li likewise needed to refine their calculations, and to consider other mirror shapes that could bounce the light back and forth. For the moment, they kept the details within Bell. They didn't want anyone to scoop them, and were particularly worried about TRG. When Ben Senitzky called Bennett to talk about gas discharge lasers, Bennett said little about his progress.

With gain demonstrated, Herriott assumed a crucial role in Bell's helium-neon laser project. For the helium-neon laser to oscillate, the light amplification in the laser cavity would have to exceed the light lost. Javan and Bennett were working to increase the gain. Herriott's job was to reduce the loss by making a nearly perfect optical cavity. The mirrors had to reflect nearly all the light that hit them, except the small amount intended to emerge in the beam. The mirrors also had to be aligned precisely with each other lest the reflected light leak out of the tube instead of hit the opposite mirror. And the beam had to pass very efficiently from the inside of the cavity to the outside world. Achieving that optical perfection was a very demanding task.

Bell's other projects lagging behind. After he and Boyle gave up on their semiconductor project, Don Nelson searched through published papers on optical spectroscopy, and found that a compound called cesium uranium nitrate had some of the desirable properties of ruby. It absorbed light at a broad range of wavelengths, and fluoresced strongly. Yet when he drew a small crystal and illuminated it with bright light, he found that the fluorescence generated in the crystal was quickly absorbed by other atoms that had been excited by the pump lamp. Kaiser and Garrett continued hunting for materials that could emit visible light when small quantities were added to calcium fluoride.

Other projects also were moving slowly. Sorokin and Stevenson waited in a holding pattern at IBM, because the computer company wasn't equipped to grow and polish optical crystals. They turned to a small optical company that specialized in crystal growth, but producing the large and optically perfect crystals they needed took time as well as expertise.

Eli Snitzer had an easier time because American Optical had equipment for both making glass and drawing the glass into long, thin fibers. He and two colleagues began adding small amounts of various elements to the cores of fibers and looking for fluorescence. Yet the process took time, and Snitzer was finishing work on an important fiber-optic project.

Irwin Wieder arrived at Varian in March to find that management had put a freeze on buying new equipment. That meant he couldn't set up the new microwave maser research laboratory they had promised him. While Maiman packed up his lab in Culver City, Wieder cooled his heels at Varian.

None of them knew how they stood in the laser race. Unlike athletes running on a track, the scientists trying to build the first laser couldn't glance around to see who was ahead or behind. They had to wait for others to report their progress publicly or privately, and scientists are taught to be cautious about that. They are supposed to check and double-check their results to make sure all the details are right. They may informally talk among themselves. Bell Labs held internal seminars, where company scientists talked freely, asking probing questions and helping each other refine their work. Only after passing through that gauntlet did Bell researchers submit papers to journals, where other scientists working in the field review them to see if they are worthy of publication. That process helped Bell papers pass muster at *Physical Review Letters*, although outsiders complained that Bell also had an inside track at the journal.

By the spring of 1960, little laser-related research had reached Samuel Goudsmit's stringent threshold of publishability for *Physical Review Letters*. Schawlow and Townes had fired the starting gun with their paper on laser theory. Their patent covered the same ideas, and Gould's patent application was tied up in litigation and classification. Javan and Sanders had proposed new types of lasers, but had not yet experimentally validated their ideas. Wieder and Schawlow had published on the properties of ruby. Maiman was about to contradict them in his forthcoming paper, but he was careful not to say where he was going with his ideas. Bennett and Javan had just seen gain, but wanted to confirm it and make sure they could build a laser. Fox and Li had to complete more calculations before publishing their results.

None of them had seen a startling claim in a 1959 Russian-language book. Valentin Fabrikant had conducted a series of experiments on mixtures of mercury and hydrogen with Fatima Butayeva, a colleague at the Moscow Power Engineering Institute. They picked mercury because it fluoresces at a host of visible and ultraviolet wavelengths in electric discharges. They added hydrogen to make excited mercury atoms drop out of the lower energy level faster, and measured what they thought was amplification of about ten percent on green and blue wavelengths of mercury atoms in a tube 36 centimeters long.

No claims like that had been made in America, and it took a long time for word to trickle through the Iron Curtain. Only a handful of the top Russian journals were translated into English. Other Russian publications rarely made it to America in any form, and few physicists outside of the Soviet Union could read Russian. That was just as well. Subtle problems with their instruments had fooled Fabrikant and Butayeva. Their Russian paper remained unknown in the West until after the first laser was demonstrated. Bela Lengyel of Hughes, a Hungarian immigrant fluent in Russian, got a copy and translated it into English. He gave a copy to John Sanders, who had his student Colin Webb try to repeat the experiment. Webb saw no amplification and concluded Fabrikant and Butayeva had misinterpreted their measurements.

15
TRIUMPH IN THE PALACE OF SCIENCE

THE NEW HUGHES RESEARCH LABORATORIES building in Malibu seemed "the palace of science" to physicist Bob Hellwarth who worked in Lyons' department and moved in along with Maiman. Technology giants across the country were creating corporate research laboratories modeled loosely after Bell Labs, but the usual results were sprawling suburban office buildings surrounded by parking lots and landscaped green space—pleasant, but routine. Very little that involved Howard Hughes was routine.

The hillside Malibu complex, with its movie-star's view of the Pacific, had been built for an east-coast electronics firm called Potter Aeronautical, which had planned to move west but changed its mind at the last minute. Del Webb, the contractor who built it, was a friend and business partner of Howard Hughes, and when the wheeling and dealing was done it was Hughes who owned the empty shell. He leased the shell to Hughes Aircraft, and Webb's construction crews began turning it into an elegant laboratory. The reclusive billionaire himself was nowhere to be seen, although rumor held that he visited the place in the middle of the night, when no one else was there. The sole owner of Hughes Aircraft was an urban myth to his employees; no one had seen him in person, but many said they knew someone who had.

Contractors laid wood parquet floors throughout the building and divided the interior into rooms. Long, narrow offices lined the outer walls, with windows looking over rolling hills and the ocean, ideal for pondering new ideas during scenery breaks. Windowless labs were on the other side of the hall, in the core of the building. In April, Maiman moved into an office facing the ocean. Asawa and D'Haenens shared the next office. On the other side, Lyons had a larger office that came with his rank of department manager. Maiman looked out the window; no one with vision could resist the view. But his heart and mind were focused on the laboratory across the hall. He was itching to get to work on the laser.

Moving his household and laboratory stalled Maiman's progress by at least three weeks. Packing and reassembling laboratory equipment takes as much care as handling fragile heirlooms. Instruments and references he needed were inaccessible while movers hauled boxes from Culver City to Malibu. The company shuffled its support infrastructure at the same time, from departmental secretaries to machine shops and lab supplies. Maiman couldn't work efficiently in the mist of the move, but he was able to write his paper on ruby fluorescence, and to do some design work. Home offered no respite with a resident toddler and a personal move in progress.

Intense and competitive, Maiman focused as tightly on his quest as a laser beam. "He was just single-minded in his devotion to this thing," recalls Hellwarth. He didn't pause to attend late-afternoon lectures that celebrated Caltech physicist Richard Feynman gave once a week at Malibu, although they were held just a short walk from his office. His obsession made him seem secretive to managers, who like Maiman, were busy adjusting to the new environment. Maiman didn't know Bennett and Javan had measured the first gain in a helium–neon mixture, but he had plenty of other reasons to worry about competition from Bell. Ten times larger than Hughes Research, and with ten times more money, Bell Labs was the odds-on favorite.

Maiman had made a compelling case for ruby, but there was no escaping doubts. Outside Hughes, Schawlow's conventional wisdom that ruby wouldn't work prevailed. Even inside Hughes, Maiman felt a lingering air of skepticism from management. Al Clogston from Bell reinforced doubts when he visited the spanking-new Malibu lab in April. Maiman recalls Clogston warning Hughes, "We have thoroughly checked out ruby as a laser candidate. It's not workable," although Clogston doesn't remember the incident. Maiman worried that Bell might know something he didn't, but he wasn't about to stop. He might be headstrong and stubborn, but he had solid evidence that ruby was a far better material than anyone outside of Hughes suspected.

The task before Maiman was a different one than faced Bennett and Javan. Their excited mixture of helium and neon produced only feeble stimulated emission, at an infrared wavelength invisible to the human eye. They needed to tweak the gain upward in every way they could, and to design optics that could capture every bit of that stimulated emission to produce a detectable beam.

The choice of ruby gave Maiman some big advantages. He could buy the flashlamp needed to excite the chromium atoms from a standard catalog. His earlier experiments also had shown that ruby converted that excitation energy efficiently into red fluorescence. That meant he wouldn't need to struggle for months to measure a small amount of amplification, as Bennett and Javan had done. Nor would he need a resident optical wizard like Don Herriott or Oliver Heavens to make the perfect set of optics that when perfectly adjusted would extract a feeble beam of stimulated emission invisible to the human eye. Maiman fully expected to see the red ruby beam from his laser.

What Maiman needed instead was a way to concentrate the lamp's intense flashes of white light onto the laser rod, and a way to make sure the red light from the rod was stimulated emission rather than ruby's ordinary red fluorescence.

When he first thought of exciting ruby with a bright continuous lamp, Maiman had planned to put the lamp at one focus of an elliptical cylinder and a ruby rod at the other. That approach didn't look promising for the coiled flashlamps, so he looked for another focusing arrangement. Inspiration came from a salesman, who said the biggest of the three coiled lamps was so intense it could ignite a piece of steel wool placed next to it. Maiman realized he didn't need special optics to focus the light. He could simply slip the ruby crystal inside the spring-shaped lamp, where the coils of the lamp would surround the rod with a bright surface (fig. 15.1).

After making that decision, Maiman designed a simple apparatus to test his laser ideas. His calculations showed that the smallest of the three lamps should suffice to drive a ruby laser, so he started with the FT-506, which could fit in the palm of his hand. He needed a small ruby rod to fit inside, so he picked one ⅜ inch in diameter and ¾ inch long—a cylinder roughly the size of the tip of an adult's finger.

He had the Hughes shop polish the ends of the ruby rod flat, perpendicular to the length of the rod and parallel to each other, although they didn't end up

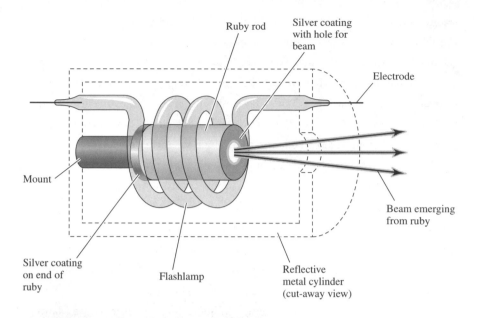

Ruby rod

Silver coating with hole for beam

Electrode

Mount

Silver coating on end of ruby

Flashlamp

Reflective metal cylinder (cut-away view)

Beam emerging from ruby

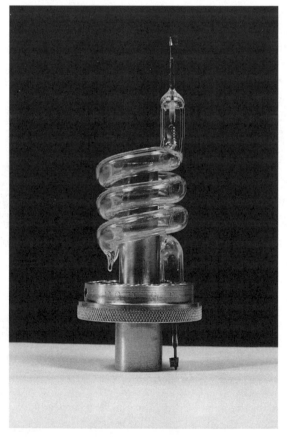

FIGURE 15.1. Structure of Maiman's first laser (top), shown with photo of the actual laser (left). (Photo courtesy of Hughes Research Labs/HRL)

perfectly parallel because the shop wasn't used to working with optics. Then he had both ends coated with silver, the most reflective metal available. Silver tarnishes readily, but he sought to keep the surface in contact with the ruby clean by having thick layers of the metal applied. He scraped the silver off the center of one end of the rod, leaving a transparent opening so the beam could escape from the reflective cavity.

Maiman had an aluminum cylinder machined and polished to slip around the spring-shaped flashlamp. The shiny inside of the tube reflected light emerging from the outside of the lamp coils back toward the ruby rod. The lamp coil absorbed some of the light, but some passed between the widely spaced coils, delivering more pump-light to the rod. Plugs that slipped into the ends of the aluminum cylinder held the ends of the lamp in place, and served as mounting points for the ruby rod and the electrode that triggered the lamp to fire. One plug had an opening for the beam to emerge. The whole package was the size and shape of a small water glass. Some light from the flashlamp leaked out through the beam opening, but the cylinder confined most of the lamp's blindingly bright flash, so it wouldn't completely dazzle the eyes of the experimenters. A separate power supply fired electrical pulses, which electrical cables delivered to the lamp.

The ruby laser design was remarkably simple compared to the one Herriott, Bennett, and Javan were developing for Bell's helium–neon gas laser. Maiman's calculations indicated he had a generous margin for error. He didn't need the best possible mirrors; simply silvering the ends of the ruby cylinder looked good enough for a first trial. Nor did he need the best possible crystal, which was a good thing because the optical quality of his ruby rod was not very good. The aluminum cylinder wouldn't reflect all the light back onto the rod, but he calculated it should reflect enough. The Hughes machine shop could make the parts that he couldn't buy. The packaging showed a good engineer's understanding of the importance of practicality. Starting with the smallest of the lamps, Maiman hoped to avoid the need for cryogenic cooling to remove excess energy. If it didn't work the first time, he could try better mirrors or bigger lamps or both.

In contrast, Bell Labs knew they were working at the margins of possibility. They had to have the best possible mirrors because they needed every possible advantage to push the helium–neon laser over the threshold of operation. They didn't see any reason to seal their experimental laser tube in a case. They didn't expect a blinding glare from the laser tube, and leaving it open to the air would help dissipate the heat. Besides, they saw it as a physics experiment and wanted to be able to adjust parameters when needed. Maiman's laser was also an experiment, but his model had convinced him that it had to work.

A critical part of the experiment was collecting conclusive evidence of its results. That was going to be important because doubters like Schawlow had insisted for months that ruby was hopeless as a laser material.

It's easy to assess an experiment if you know how it should behave. The test to see if you've wired an electric light properly in your house is simple—put in a bulb and turn on the power. If the bulb goes on and stays on, it works. The same principle applies if you add a new memory chip or a new hard drive to a computer. Once you put the pieces back together and switch on the power, the computer should recognize the new memory or drive. That moment of truth tests your success.

Nobody had made a laser before, so Maiman had to devise his own set of tests. It wasn't enough to look for light; ruby glowed red with fluorescence when illuminated by violet, blue, or green light. He needed a way to distinguish clearly between ruby's ordinary red fluorescence and the coherent beam generated by the amplification of stimulated emission in a resonant cavity. Fortunately, theory predicted some important differences.

The most obvious difference was that laser light should be concentrated in a narrow beam; the idea that had excited Gordon Gould and others. Simple fluorescence is spontaneous emission, so it should emerge from a sample of ruby in all directions. Stimulated emission that amplified light bouncing back and forth between a pair of mirrors should be directed along a line between the mirrors.

Yet it wasn't that simple for a ruby rod. If the entire volume of the rod glowed uniformly and none of that fluorescence was absorbed inside the ruby, the light would look brightest when you looked at the thickest parts of the ruby. Thus more light would emerge from the ends of the rod than from the sides. Maiman's rod was short and stubby, but the package he had designed let him observe only the light coming from the end of the rod. He hoped to see the light narrow into a beam, but he didn't expect it to be very tightly focused. He also knew that merely projecting a red spot onto a wall wouldn't be enough convince skeptics. He needed additional quantitative measurements, which other physicists could reproduce in their own laboratories with their own quantitative instruments. That reproducibility would be the acid test.

A more subtle effect, detectable only with fast electronic measurement instruments, was how quickly the intensity of the red ruby emission dropped after the pulse from the flashlamp ended. One attraction of ruby for use in a laser was that chromium atoms retained energy they absorbed for a long time—atomically speaking. Chromium atoms quickly released some of the energy they absorbed from the flashlamp, dropping into a metastable state. If left alone, they spontaneously emitted the rest of the energy as red fluorescence in about three milliseconds. However, if other photons with the right energy came along during the milliseconds that the chromium atoms retained that extra energy, they could stimulate the atoms to emit their extra energy before they dropped to the lower laser level on their own.

As long as the ruby produced only spontaneous emission—ordinary fluorescence—the intensity of the red light should drop by a factor of two every three milliseconds. If laser action switched on, stimulated emission would drain the light energy faster, producing a shorter, more intense spike of light than ordinary fluorescence. The human eye can't respond fast enough to sense that difference, but sensitive electronic instruments can measure it. Opto-electronic detectors convert light into an electric signal that reproduces the rise and fall of the optical pulse, and an oscilloscope can trace the rise and fall of that electronic pulse on its screen for the human eye to see. When laser action began and stimulated emission overwhelmed spontaneous emission, Maiman expected the peak of the pulse on the oscilloscope screen to become shorter and sharper, followed by a rapid decline as shown in fig. 15.2. That was one test for a working laser.

His second test for a working laser was even more subtle, but like the first it depended on the physics of stimulated emission. He expected stimulated emission from ruby to span a much narrower range of wavelengths than spontaneous emission. Ruby fluoresces across a range of wavelengths that all look quite red to the eye, but they are not exactly the same. The fluorescence is most likely to be at the center of the range, with the probability dropping off at the sides, so the light intensity varies in the same way. Stimulated emission spans the same range of wavelengths as the fluorescence, but it amplifies the initial fluorescence. That amplification process concentrates stimulated emission in a much narrower range of wavelengths, as if you multiplied the emission curve by itself many times. The gain is highest in the middle of the curve, so that wavelength is most likely to be amplified. Wavelengths to the side are less likely to be amplified. If the fluorescence intensity differs by a factor of two, the difference will double at each round of stimulated emission. Laser emission should span a much narrower range of wavelengths than fluorescence, with the exact variation depending on

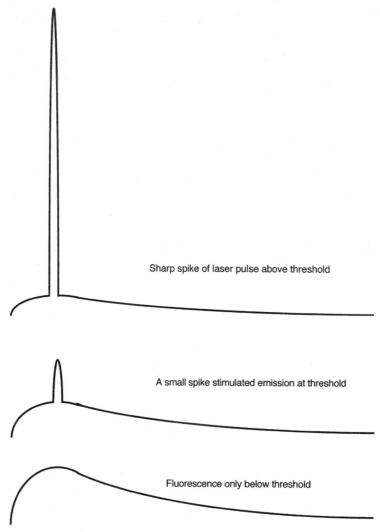

Sharp spike of laser pulse above threshold

A small spike stimulated emission at threshold

Fluorescence only below threshold

FIGURE 15.2. Rudy laser pulses rise and fall much faster than fluorescence produced by lower-power illumination. When he started with modest flashlamp pulses, Maiman saw only the rise and slow fluorescence, at bottom. At laser threshold, a small spike of stimulated emission. When the flashlamp power exceeded the laser threshold, stimulated emission extracted energy from the ruby in a brief burst much shorter and more intense than the fluorescence he saw at lower flashlamp powers.

the nature of the laser material and the resonant cavity. Maiman expected a good spectrometer to show this line-narrowing effect, shrinking the band of red fluorescence narrow to a laser line.

Both pulse-shortening and line-narrowing are closely related to the existence of a distinct threshold for laser action. When input power is low, nothing happens beyond the emission of a little fluorescence, which increases a little as the power increases. When the input power increases above a threshold value, stimulated emission takes over and the laser turns on, with power increasing much more rapidly. The effect is like igniting a flame, or pushing an object along an increasing downhill slope until it finally starts moving on its own.

Laser threshold is the point where the amount of light added by stimulated emission exceeds the light lost for each round trip the light makes through the laser cavity. The process is a little like trying to fill a leaky bucket. Each round trip, spontaneous emission produces some photons, stimulated emission produces some, and some photons leak out. Below threshold, virtually all of the light is spontaneous emission, which increases with input power but is not enough to make up for the light that leaks out. Stimulated emission becomes significant once the pump energy produces a population inversion, but a population inversion is not enough to reach laser threshold. Only when the light gained from stimulated emission equals the light lost internally does the laser reach threshold. For example, if one percent of the light is lost on each round trip of the laser cavity, and another one percent emerges as the beam, the gain on a round trip of the laser cavity must be two percent to reach laser threshold.

Once you exceed laser threshold, the light power increases sharply and the optical bucket starts to fill. If the increase in power—the gain—is only one percent per pass, the power increases 64.4 percent after 50 round trips and cascades upward from there. After 100 round trips the power is 170 percent (a factor of 2.7) higher, and after 500 round trips the power increases by a factor of 145. Since the round trips are at the speed of light, we see a nearly instantaneous jump in power as the laser exceeds threshold and turns on. The power doesn't grow infinitely because only a limited number of atoms can be excited at any instant, and the amplification process saturates.

Maiman's experimental plan was to slowly crank up the power applied to the flashlamp and watch what happened. He expected to see ordinary red fluorescence at lower power, increasing gradually until he reached threshold. Above threshold, the laser would turn on, emitting a much brighter red beam. Crossing laser threshold was important, but he knew that alone would not convince skeptical physicists. He wanted to record both pulse-shortening and line-narrowing on his instruments.

With the experiment designed, D'Haenens and Maiman unpacked and assembled their equipment in their new garage-sized lab as spring greened the Malibu hills. Everything was ready to go for their first test of the ruby laser on the afternoon of May 16, 1960. Electrical cables connected a power supply to the two ends of their flashlamp. Another cable linked the trigger electrode to a source of high voltage pulses synchronized with an oscilloscope. The heart of the power supply was a big capacitor, which slowly accumulated a charge and held it until the trigger voltage ionized the xenon in the lamp and allowed the capacitor to discharge its energy through the lamp.

Linking the trigger voltage to the oscilloscope synchronized the firing of the lamp with the trace drawn across the oscilloscope screen, so they could compare the sequence of events for many different pulses, as they changed the voltage across the lamp. The oscilloscope trace measured the amount of red light registered by their sensor. A point swept across the screen at a constant rate, rising when the detector sensed light. Horizontal markings on the screen measured the passage of time, and vertical markings measured voltage from the detector. They set the trigger voltage to fire at a certain point in the cycle. A memory inside the oscilloscope recorded the traces from successive pulses, an unusual feature in 1960, but one that made it much easier to study pulses lasting just a few milliseconds.

They aligned the little laser package to focus the ruby output into their monochromator, which spread the light out into a spectrum, then directed a thin slice of that spectrum to a sensitive light detector called a photomultiplier tube, isolating the red ruby emission. Photons entering the photomultiplier triggered a cascade of electrons, producing a strong electrical signal that it delivered to the oscilloscope. The oscilloscope trace rose and fell with the strength of the signal, measuring how much light reached the tube. Each sweep showed how emission from the ruby varied as the pulse of light from the flashlamp hit it.

It was a well-designed experiment, but it was not a make-or-break moment. Maiman and D'Haenens had done their careful best to make sure the ruby laser would work, but it was just their first cut. What mattered most was that they learned something from the results and made progress toward their ultimate goal of a working laser. Nobody was peering over their shoulders. If something went wrong, they could go back and try again. Maiman had bigger flashlamps in reserve, in case the little one didn't have enough power to push ruby above the laser threshold.

They started firing the flashlamp with pulses of 500 volts, a modest level for their lamp, and well below the threshold for laser action. When they fired the lamp, the lamp flashed, and the oscilloscope showed a trace of light rising in power, then dropped in about three milliseconds, just what they expected from ruby fluorescence. They adjusted the knobs and started turning up the voltage step by step, expecting the power in the ruby pulse to increase a little bit with each step.

The experiment was simple and repetitive. Turn up the voltage, fire the pulse, and look at the oscilloscope trace. Today it would be computer controlled, but in 1960 Maiman and D'Haenens did it by hand, firing a single pulse that triggered both the flashlamp and the oscilloscope scan. Gradually the power level increased, but the more powerful pulses dropped with the same three millisecond decay time. The little aluminum cylinder contained most of the light from the flashlamp, but bright flashes of ruby fluorescence scattered red light through the room.

Maiman and D'Haenens kept their eyes on their instruments, but the red flashes inevitably reached their eyes. They were bright enough to bleach the red color receptors in Maiman's eyes, so he couldn't see the red flashes well. They had little effect on the color-blind D'Haenens. He and Asawa had worked out the nature of his impairment during earlier ruby experiments. D'Haenens had the normal number of blue and green color receptors, but had very few red receptors, so his eyes barely responded to red light.

The brightness of the ruby pulses increased steadily as they turned up the voltage. Their instruments showed the increase was proportional to the extra energy delivered to the lamp, a sign that everything was working properly. They kept turning the voltage up, and watching the instruments carefully each time they fired the flashlamp. The oscilloscope trace was a heartbeat on the screen. The flash brightness increased in small steps until they turned the power supply above 950 volts. Maiman saw the oscilloscope trace surge. "The output trace started to shoot up in peak intensity and the initial decay time rapidly decreased. Voilá. This was it! The laser was born!" he wrote. The pattern was just what he had expected, shown at the top of figure 15.2.

As the laser surged to life, its brilliant red glow permeated the room. The two men had aimed the little laser cylinder at a piece of white poster board, and Maiman's red-dazzled eyes were focused on the instruments telling the story of his success. When the laser went above threshold, D'Haenens saw its red light for the first time. The beam formed a red horseshoe-shaped spot a few degrees wide on the poster board, so bright that even the few red sensors in D'Haenens' eyes could see it. He jumped with joy. Maiman felt numb, relieved that his gamble had paid off.

They called in Charlie Asawa and Ray Hoskins, another physicist in the group, whose office was across the hall, and basked in the red light of glory.

Word of the breakthrough spread quickly, and others in the atomic physics department came to take a look. Bob Hellwarth, who had brainstormed with Maiman on how to recognize laser action, congratulated him. Harold Lyons was delighted. His department had scored a breakthrough, beating Bell Labs to a prize that Bell had spent far more time and money seeking.

The next morning, Lyons showed up bright and early at Maiman's office, with plans to put out a press release. Lyons knew the importance of publicity; he had gotten considerable ink himself in January 1949 when he unveiled the world's first atomic clock at the National Bureau of Standards. Lyons had dressed up the otherwise dull racks of equipment by gold-plating the 30-foot tube containing the ammonia molecules, which absorbed microwaves at 23.87 gigahertz, and coiling it around an ordinary electric clock he had mounted above the real innards of the clock. The press had loved it (see fig. 7.1, p. 85).

Maiman was not ready to go that far yet. He was proud of his achievement, but he also insisted on precision. His first experiment had left some doubts in his mind. He had not seen as sharp a jump in laser power as he had expected at laser threshold. He wanted to track down the cause of that anomaly before raising an embarrassing false alarm. He suspected the cause was imperfections in the ruby crystal, which also spread the beam out on the poster board, and persuaded Lyons to hold off until he could make additional measurements to verify laser action.

Maiman immediately ordered three new ruby crystals, and asked the supplier to cut them into rods and polish their ends, because he wasn't sure the Hughes shop could handle the hard ruby crystals properly. Asawa soon found that the first ruby rod had internal flaws that caused light scattering. In addition, the end faces had not been made precisely parallel to each other, so the light did not bounce back and forth properly between them. However, only a single company, a division of Union Carbide, could supply the ruby crystals, and Maiman would have to wait weeks for delivery. For the moment he had to do the best he could with the original.

In the first experiments, Maiman and D'Haenens had concentrated on the easiest measurement, the change in how quickly the red emission dropped. The decrease they saw was consistent with laser action, but Maiman wanted more evidence. He wanted to show that the range of wavelengths in the red ruby output narrowed when stimulated emission dominated above laser threshold. He also wanted to verify a subtle effect that he had predicted. Pink ruby fluoresces

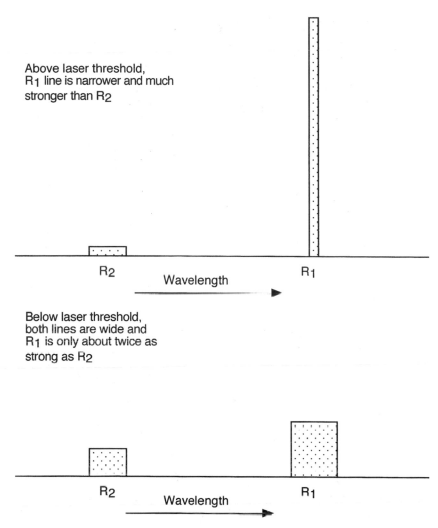

Above laser threshold,
R₁ line is narrower and much
stronger than R2

R2

Wavelength ▶

R1

Below laser threshold,
both lines are wide and
R₁ is only about twice as
strong as R2

R2

Wavelength ▶

R1

FIGURE 15.3. Laser operation narrowed the range of wavelengths and shifted the balance between the two red lines of ruby. Spontaneous emission from ruby (at bottom) was fairly even divided between two lines called R1 and R2. Above laser threshold, laser oscillation narrowed the R1 line and greatly increased its power, leaving the R2 line weak. Maiman had predicted this effect, and he considered observing it proof of laser operation.

at two closely spaced red lines, called R_1 and R_2, at wavelengths of 694.3 and 692.9 nanometers, respectively, at room temperature. This happens because the metastable upper laser level is actually a pair of closely spaced energy states. When the chromium atom is in the lower state, it emits on the R_1 line; when it is in the upper one, it emits at the shorter-wavelength R2 line. When ruby fluoresces, it emits about twice as much light at the longer R_1 wavelength, and Maiman expected stimulated emission to increase the dominance of the R_1 line.

Maiman's modest monochromator couldn't make the required measurements. He needed an expensive high-resolution spectrograph. Another Hughes physicist, Ken Wickersheim, had just received exactly the instrument that Maiman needed, but he had a backlog of experiments because he had been waiting for its arrival for six months. At Maiman's request, Lyons pulled rank and commandeered the new instrument so Maiman and Asawa could perform additional tests to verify that their ruby really was working as a laser. Wickersheim, an old classmate of Asawa's, was furious and took off for an extended camping vacation in the Sierra Nevadas to get away from Hughes and cool off.

Maiman and Asawa quickly got results. When Asawa ran the ruby output through the new spectrograph, he recorded the line narrowing that Maiman had predicted (fig. 15.3). The light was so bright that it initially blackened the photographic plates they used to record the spectrum. Asawa had to insert high-density filters or spread the light over a larger area to reduce the light level enough for the plates to record the spectrum properly. The spectrograph also clearly separated the two closely spaced red lines. Asawa found that the R_1 line was at least 50 times brighter than the R_2 line, confirming another of Maiman's predictions. Maiman and Lyons were delighted; they had results good enough to publish, and a big success to report.

Little word of Maiman's work went up the Hughes chain of command until the second round of experiments. Maiman "was very secretive," until he began to get good results, recalled George Smith, who managed another department and was on the same level as Lyons. Smith didn't hear a thing right after Maiman fired the first laser pulse in mid-May. But once the second experiment convinced Maiman he had the laser working properly, word spread around the lab in a matter of hours.

16

AN UNEXPECTED STRUGGLE FOR ACCEPTANCE

THE HARD PART SHOULD HAVE BEEN OVER. After nine months of intense effort, Maiman had reached his goal. With data in hand, Maiman sat down to write a paper reporting his evidence for the ruby laser. The laser race was too tight to wait for arrival of better ruby crystals. His manuscript passed swiftly through the Hughes chain of command, but the company's patent department didn't see anything worth filing—later costing Hughes any shot at overseas patents. On June 22, 1960, Maiman air-mailed it to *Physical Review Letters*, the country's leading journal for reporting hot research in physics.

The journal's high visibility and rapid publication time made it a logical choice for an important paper Maiman wanted to get into print as soon as possible to stake his claim of success. He had every reason to be confident of acceptance. He had checked and double-checked his experimental results to his own exacting standards. He and his managers had checked his manuscript carefully. The journal had just published his research on ruby fluorescence in its June 1 issue. The laser was clearly a more important discovery, and he expected the journal to accept it with minimal fuss and have it in print in weeks.

However, Maiman had not reckoned with the feisty Dutch physicist who had founded *Physical Review Letters* in 1958 for quick publication of short "letters" that had been getting lost among the longer articles of the venerable *Physical*

Review. Samuel Goudsmit (fig. 16.1), then 57, had won a series of awards for discovering electron spin as a graduate student at the University of Leiden in 1925. A senior scientist at the Brookhaven National Laboratory on Long Island, he was also editor-in-chief for the American Physical Society, shepherding the society's scholarly journals from his Brookhaven office.

Trained in the academic tradition of "pure" physics, Goudsmit had the attitude that fundamental physics was more important than applications. It was a common attitude among prominent members of the physics establishment, and Goudsmit's eminence and central position at the physical society gave him serious clout. "He had strong opinions, and was not afraid to voice them," recalls Ben Bederson, who later became editor-in-chief at the American Physical Society. Younger physicists found him to be "a rather domineering figure." Goudsmit had little patience for nonsense and a broad definition of what constituted it.

FIGURE 16.1. Samuel Goudsmit, editor of *Physical Review Letters* (Courtesy of AIP Emilio Segre Visual Archives, Goudsmit collection)

He had succeeded in attracting a growing tide of submissions to his new journal, but he was not pleased that many of them described practical devices. Microwave masers were hot, so they became a particular sore point. In an editorial titled "Masers" in the August 1, 1959, issue, Goudsmit decreed that maser physics had become routine. "[M]any of the maser papers we receive contain primarily advances of an applied or technical character and comparatively little physics, and are thus more suitable for other journals," he wrote. In the future he would consider "those few [maser papers] which contain significant contributions to basic physics." Anything else should go elsewhere.

Goudsmit had a point. Microwave masers were maturing. The basic concepts were well-established, and most activity was devoted to building more practical devices. The American Physical Society published the *Journal of Applied Physics* to carry such papers, and that's where Goudsmit wanted all "maser" papers to go. Maiman had described his dry-ice-cooled microwave maser in its pages.

Maiman rightly considered the optical laser to be fundamentally different, a whole new class of device that was an important breakthrough. Yet in a misguided attempt to ease the paper into print, he chose to describe it in Townes's words as an "optical maser." Goudsmit was not in an open-minded mood on June 24 when he saw Maiman's manuscript titled "Optical Maser Action in Ruby." He clipped a copy of his no-more-masers editorial to the manuscript, and immediately returned it along with what Maiman described as "a curt rejection letter." In it, Goudsmit wrote "It would be more appropriate to submit your manuscript for possible publication to an applied physics journal, where it would receive a more appreciative audience."

A second issue also contributed to the rejection. Another of Goudsmit's pet peeves was physicists who reported a series of minor advances in separate papers, a sin called "serial publication." He had fumed about such papers flooding his desk in another editorial. Maiman's paper on ruby fluorescence had just appeared in the June 1 *Physical Review Letters*, under the title "Optical and Microwave-optical Experiments in Ruby." The similarity must have caught Goudsmit's eye. He later told Townes and Schawlow that he had rejected Maiman's laser paper because the journal had just published his report on ruby fluorescence.

Goudsmit obviously didn't understand what Maiman had done. He may not have read anything but the title. His summary rejection bypassed the normal process of refereeing papers. Editors of scholarly journals are supposed to function as judges, referring papers to specialists who function as a jury, reviewing papers in their specialties. The goal is to have expert referees review the papers, to identify any mistakes or misconceptions, and to sort the trivia from the important advances. Referees often request clarification, and if the authors satisfy their questions, the papers are published. Goudsmit didn't give Maiman a chance, acting as judge and jury. Some observers have suggested that Bell Labs might have deliberately torpedoed Maiman's paper, but no evidence supports those claims. Goudsmit blew it all by himself.

Maiman still steams about the rejection after 40 years. His anger is justifiable. He had followed all the rules to challenge the establishment view that ruby wouldn't work. His careful experiments showed that ruby worked as a laser. Others would soon prove that the experiments were repeatable. Yet the establishment turned their backs on him anyway.

In part, Maiman blames his own effort to please the establishment. He used the term "optical maser" reluctantly, "because I knew the editorial staff at *Physical Review* was very stuffy." Hughes physicists preferred "laser," but that was Gould's word. Townes and Bell Labs insisted on "optical maser," and Maiman knew they were well connected. Worried that editors would consider a "laser" as just another device to be ignored, rather than a new physical principle, Maiman tried to play it safe by using the establishment's term "optical maser."

Yet the real problem was Goudsmit. Maiman pleaded for reconsideration, but to no avail. Peter Franken called *Physical Review Letters* to ask what had happened. Goudsmit was on vacation, but his assistant, George Trigg, told Franken that Goudsmit had rejected the paper. After talking with Maiman, Franken called back to complain, and an apologetic Trigg promised to pass the complaint along to Goudsmit. Nothing happened. Under Goudsmit, there was no formal appeal process; once Goudsmit made the decision, it was all over. Franken argued long and loud, chain smoking and pacing as he held the only phone in his University of Michigan laboratory. He pushed the editors to admit they had made a mistake, but that was not to be. No mere junior faculty member or industrial physicist could get away with telling Samuel Goudsmit he was wrong. Maiman received a form letter informing him that the journal never reconsidered papers that had been rejected for any reason. The editors of *Physical Review Letters* had donned the cloak of infallibility and slammed the door in Maiman's face. It was a stinging rejection that could only amplify Maiman's fears that the establishment was determined to ignore him. Expecting a fair hearing, he had been railroaded into a kangaroo court.

Lyons and other Hughes managers were as eager as Maiman to get the news out. The rejection threatened Maiman's claim to have won the race to build a laser. Hughes was keenly aware that Bell and TRG had massive, well-funded laser programs. They worried that either one could snatch credit for the laser by getting a paper into print first. *Physical Review Letters* had been the first choice because it could publish papers within weeks of receiving them. For most other journals the delay was months, and Hughes knew the clock was ticking.

The second choice was the widely read British weekly *Nature*, which could get short letters into print very quickly. Maiman dashed off 300 words reporting that he had stimulated the emission of visible light and air-mailed it to London. The paper provided little detail on the laser itself, but *Nature* accepted it immediately, scheduling it for the August 6, 1960, issue as "Stimulated Optical Radiation in Ruby." Once it was clear that *Physical Review Letters* would not publish anything, Maiman submitted his more detailed description of the laser's workings to the *Journal of Applied Physics*. Like *Physical Review*, it was published by the American Physical Society, but it had no hang-ups about applied physics and quickly accepted Maiman's report. However, publication would take six months.

Hughes was already thinking about a press release in late June, when Malcolm Stitch, another Hughes physicist, placed an urgent call to Malibu from a conference at the University of Rochester. He had just heard Oliver Heavens tell the meeting that Columbia was close to success with the cesium laser.

A bright but volatile character, Stitch was on his third marriage, and often in hot water with Hughes managers. He didn't realize that Heavens was not in close touch with Abella and Cummins, who were having lots of trouble with cesium. Peter Franken, himself a playful spirit, thought Heavens was kidding when he said that Columbia was nearly ready to report a cesium laser. If he was, the joke went right past Stitch.

Stitch couldn't keep his mouth shut about Maiman's laser when he talked with Bell Labs scientists at the Rochester meeting. They took the news back to New Jersey, where Schawlow recalled it caused "both excitement and puzzlement." With no official confirmation, Bell treated Stitch's hint as an unsubstantiated rumor. Meanwhile, Stitch, who enjoyed playing reporter, was more discreet in a front-page article he wrote for the weekly *Electronics News* about the Rochester conference. Instead of lasers, he wrote about new fiber-optic research that Eli Snitzer had described at the meeting—much to the surprise of Snitzer, who hadn't expected his report to get much press attention. In retrospect, it looks like Hughes may have persuaded Stitch to keep the lid on lasers for a little while.

Stitch's call prodded Hughes management. The *Nature* letter was safely in the works, but the *Journal of Applied Physics* paper was months away. They decided to schedule a press conference and announce that Hughes Research Laboratories had won the laser race before someone else tried to claim the prize.

It was an important milestone for the company. Hughes Aircraft was a successful corporation, and the research labs were well regarded in the Pentagon. Yet none of its previous inventions had headline quality. The laser gave Hughes a chance to step into the big leagues of corporate research. It gave the upstart Californians a chance to tweak the noses of the east coast physics establishment, centered on Bell Labs, Brookhaven, Columbia, MIT, and Harvard. No longer would Hughes be just another aerospace contractor, best-known for its eccentric namesake and sole owner.

The company had a contract with a top-level public relations firm, Carl Byoir & Associates of Los Angeles, which handled the affairs of Howard Hughes himself. They set the press conference for July 7 at the Delmonico Hotel in Manhattan—a high-profile site to attract the cream of the media. Lyons helped write the press release, coming up with the odd description of a laser as "atomic radio light." Lyons also decided that he should be the one flying to New York to handle the press conference because he had experience with the media that Maiman lacked.

Lyons had demonstrated his knack for promotion with the atomic clock. Maiman, at age 32, was four years younger than Lyons had been at the time of his discovery, and Lyons may have thought that was too young. He may have picked up the idea that management deserved a large share of the glory for successful projects at the National Bureau of Standards. NBS director Edward U. Condon had appeared with Lyons to introduce the atomic clock, and if his expression in figure 7.1 is any indication, Lyons probably was not very happy about it.

Yet Condon had played a different role with the atomic clock than Lyons had with the laser. Condon was a nationally prominent physicist and head of the whole bureau; Lyons was a department manager. Condon probably relished the favorable publicity after surviving a bruising attack by the House Un-American Activities Committee, but he didn't try to cut Lyons out of the picture. The atomic clock photo was a traditional one of a top executive congratulating a worker for a job well done.

The plans for the laser press conference were something quite different. Lyons deserved some credit for assembling the talented people who made up the atomic physics department and for supporting the laser project, although his support sometimes had been grudging. But the project and the ideas were Maiman's, and he deserved the credit and the starring role at the press conference. Lyons's role should be on the sidelines. Whatever Lyons actually intended, his plans to run the press conference while Maiman stayed home came across as an effort to steal credit for the ruby laser so Lyons could promote his own ambitions to climb the Hughes management ladder.

Maiman had ambitions of his own, which had driven his laser project. Lyons's plans to represent the group at the press conference brought the long-simmering tensions between the two to a rapid boil. Maiman was furious, and a major confrontation ensued as he defended his claim to the laser.

Maiman was not the corporate politician that Lyons aspired to be. But Hughes drew its management from the ranks of its top engineers and physicists, and they thought the man who made the laser should be the one to tell the press about it. When Mal Stitch heard of Lyons's plans, he quickly came to Maiman's defense and complained to Malcolm Currie, who was head of the physics laboratory. Currie listened. He, George Smith, and Harley Iams formed a troika of managers who were running the lab from day to day in Andy Haeff's absence. The three solidly lined up behind Maiman and said he should be the point man at the New York press conference. Lyons was thoroughly chewed out for his actions, and soon left Hughes, under something of a cloud. Word of the incident spread, hurting Lyons's reputation. When Lyons visited Michigan afterward, Franken's students virtually shunned him.

Photos were a must for press releases, so Byoir sent a top publicity photographer to Maiman's lab. The photographer's favorite trick was to pose a device in front of the inventor's face, but the first laser was the size of a juice glass, so it was too small to work. Looking around the lab, the photographer spotted another laser set-up, with a longer ruby rod inside the mid-sized FT-503 flashlamp. That met the photographer's aesthetic test, but Maiman wanted a picture with his little first laser. He complained that showing the bigger laser instead of the first one was inaccurate, but the photographer claimed creative license. "You do the science, I do the pictures." He snapped separate shots of other equipment, but the centerpiece of the press kit was figure 16.2, showing Maiman holding the mid-sized flashlamp design, which he calls "not the first laser." It was the world's introduction to the laser and has been reproduced in countless books and articles ever since.

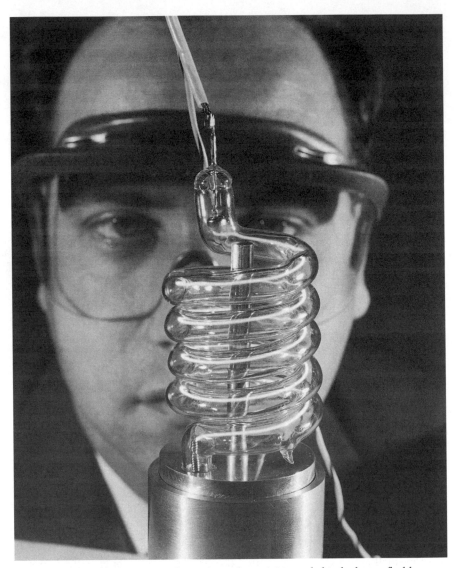

FIGURE 16.2. Hughes press-release photo shows Maiman behind a larger flashlamp with a longer ruby rod than he used. This was the laser duplicated by TRG, Bell Labs, IBM, and others. Compare with the real first laser in figure 15.1. (Publicity photo courtesy of Hughes Research Labs/HRL Laboratories LLC)

A crew from Hughes and Byoir Associates flew to New York the night before the press conference. Maiman brought the real first laser to show off, but the head of Hughes public relations complained. "It looks like something a plumber made!" The little laser may not have fit the public image of a dramatic break-through in the age of bigger is better, but its simplicity had a compelling elegance compared to refrigerator-sized contraptions like Townes and Gordon's first microwave maser.

The press release was headlined "U. S. victor in world quest of coherent light," and made a point of using the word "laser" while noting it was "sometimes called an optical maser." The coherence of light waves marching in phase was a logical starting point. Physicists and engineers previously had produced coherence only at the relatively low frequencies of radio waves and microwaves, from thousands of hertz (kilohertz) to billions of hertz (gigahertz). Maiman had leapfrogged from the top of the microwave range, tens of gigahertz, to a frequency of nearly 500,000 gigahertz. The press release revelled in the giant number of the frequency, emphasizing for public consumption the huge gap Maiman had crossed. Hughes handed out preprints of the paper Maiman had in press at the *Journal of Applied Physics*, to provide details for the more technical members of the press and to show that it was being published. Later Maiman mailed copies to some 200 people, including everyone listed as attending the Shawanga Lodge meeting.

Maiman showed his little laser proudly at the press conference, no matter what the publicists thought. The release cited its ability to generate high frequencies at high intensities despite its modest size. It also stressed that Hughes had developed the laser with its own money, not under a government contract.

It was natural to ask what a laser could do. Maiman cited five potential uses, only one of which has not gained wide acceptance in the decades that followed:

1. The first true amplification of light (now used in long-distance fiber-optic communications).
2. A tool to probe matter for basic research (laser spectroscopy, the subject of several Nobel Prizes).
3. High-power beams for space communications (which has been demonstrated but is not in regular use).
4. Increasing the number of available communication channels (fiber optic communications).
5. Concentrating light for industry, chemistry, and medicine (laser cutting, drilling, welding, and surgery).

Hughes attracted a good selection of reporters, from the *New York Times* and other newspapers, major wire services, big magazines, and specialized publications. Most of the press conference came off quite well. When the formal question-and-answer period ended, reporters swarmed around Maiman. A reporter from the *Chicago Tribune* began asking about a concept that had gone unmentioned in the press release—laser weapons. Had Hughes developed a death ray?

It was a question that would plague the laser community for decades. Scientists and engineers understood the laser beam was just a beam of light, but the popular culture was heavily influenced by pulp science fiction and B movies. From H. G. Wells to Buck Rogers, science fiction had armed its heroes and villains with ray guns and death rays. Artificial satellites, once the stuff of science fiction, now orbited the Earth. Plans were afoot to send people into space. The world lived in a balance of nuclear terror. It was no wonder a reporter would ask if this new technology might become a weapon of the atomic age.

Maiman tried to dodge the question, saying he didn't think laser weapons would be possible for at least 20 years.

The reporter came back again and again, and each time Maiman tried to avoid a direct answer. Finally, the reporter asked, "Are you willing to say that the laser could not be used as a weapon?"

Maiman was caught. As a scientist, his responsibility was to be honest and objective. He wasn't going to lie. "I can't say that," he admitted.

That was all the reporter needed, and others followed his lead. The next day John Osmundsen had a sober story leading on the bottom left corner of the front page of the *New York Times*. But wild predictions of futuristic weapons outshouted the sensible stories on the newsstands. Maiman returned home to find his achievement billed in red type two inches high on the front page of the *Los Angeles Herald*: "L.A. man discovers science fiction death ray." A Pasadena newspaper headlined, "Death ray possibilities probed by scientists," and warned that a laser-equipped satellite could rule the Earth. Even the normally sober *Newsweek* joined in the laser weapon hysteria after Princeton physicist Robert Dicke told a reporter, "This might very well be the forerunner of the death ray." The headlines were all over the newsstands when Ken Wickersheim finally hiked down from the mountains. He stopped in a small grocery store and discovered a newspaper headline screaming, "L.A. man invents light brighter than the sun." After glancing at the article, he later told Asawa, "I almost turned around to hike back into the mountains."

One reporter found the whole thing incredible and loudly refused to believe Maiman's report. Jack Jonathan, who covered science for *Time* and *Life* magazines, insisted he would write nothing about the laser until someone else had confirmed it. True to his word, *Time* did not mention the laser until Bell Labs published a paper reporting building their own ruby laser in October.

Two leading industry magazines, *Electronics* and *Aviation Week*, described the laser prominently and reasonably accurately. Popular magazines let their imaginations run riot. *Popular Science* predicted that laser transmitters for radio and television "would be a spectacular sight, flashing colors like huge, gaudy neon signs." The writer envisioned optical radars, far more precise because of their shorter wavelength, and "fantastic searchlights" that could focus a laser beam so that it spread to only 100 feet across on the way from Los Angeles to San Francisco. Inevitably, the magazine mentioned weapons. "A death ray is another possibility. The laser's beam can be focused so sharply that a fantastic amount of energy is concentrated into a pinpoint. It would certainly cook bacteria—to sterilize utensils—and maybe kill people as well."

The news made a big splash in the research community, and the days after the press conference became a rush of activity for Maiman. He flew to Washington to describe his laser to a group of military officials organized by Townes. A parade of curious scientists began visiting Hughes to see what the laser was all about. Asawa estimates well over a hundred came during the summer, including Peter Debye, a Nobel laureate famed for his studies of molecules. Yet despite the headlines, Asawa doesn't recall any reporters stopping by for a first-hand look.

Meanwhile, Maiman tightened his control over the laser project, assigning specific tasks to Asawa, D'Haenens, and others working for him, and trying to block anyone at Hughes from working independently on ruby lasers. When the public relations photographer had come to shoot pictures of Maiman and the laser, D'Haenens playfully took a Polaroid photo of himself and Asawa, labeled it ". . . and the boys who helped." He handed it to Maiman after the photo session, but Maiman was no longer one of the boys. He was shifting his role from lead research scientist to manager of the laser project. Perhaps shaken by his clash with Lyons and still assimilating the implications of his success, he seemed to become more distant.

On July 20, Maiman finally received the three small ruby rods he had ordered from Union Carbide, ground and polished to the same shape and size as his first rod. As he had hoped, the optical quality of the new rods was much better. When he put one of the new rods into his little laser he found a sharp laser threshold as he turned up the power—just what he had expected. The beam formed a tightly focused red spot on the wall, less than one-third of a degree across, which he noted at the end of July in an advance abstract he wrote for an October optics conference. That should have answered any doubts once and for all.

The Air Force quickly gave Hughes and Maiman a vote of confidence with a contract to develop bigger and better lasers for use as optical radars. Hughes also gave Maiman a promotion when Lyons left. George Birnbaum moved up to Lyons's old post as department head, and Maiman took over Birnbaum's job as section head.

Thinking his paper was securely in press at the *Journal of Applied Physics*, Maiman took the new laser rods and began a more detailed series of experiments and a thorough analysis of laser operation, with help from Ray Hoskins, D'Haenens, Asawa, and Viktor Evtuhov. However, he soon got another rude shock. The paper he had written for the *Journal of Applied Physics* on the first laser showed up in the September issue of *British Communications and Electronics*, a publication that Maiman had never heard of.

The British magazine had received a copy of the paper along with the laser press release from Byoir Associates. One of the editors, Charles Marshall, was sufficiently impressed that he decided to publish the paper in its entirety, although he didn't bother to check with Maiman, Hughes, or Byoir Associates. It was an unusual move, and Maiman was amazed when he received a letter from the editor announcing the publication.

He was also dismayed by the consequences. The British magazine was an obscure one, very hard to find in the United States. But it represented a publication, and even if it was unauthorized, once Maiman's paper had appeared in one journal, no other journal would touch it. Maiman had to withdraw the paper from the *Journal of Applied Physics*. That remarkable streak of bad luck meant that the only readily accessible record of the first laser was the scanty letter in *Nature*. Maiman's group was in the midst of detailed studies of the ruby laser that would eventually become a pair of long and comprehensive papers in *Physical Review*, but they wouldn't be submitted until January 1961, six months after the press conference.

17

"WE WERE ASTOUNDED"—
A STUNNED REACTION

RIPPLES OF REACTION began spreading as soon as the reporters left the Hughes press conference. Walter Sullivan, the science editor for the *New York Times*, didn't attend, but afterward he tried to reach Townes, whom he often asked questions about physics. Townes was away from Columbia, but the physics department announced the call over its public address system, and Abella picked it up, identifying himself as a graduate student.

When Sullivan told him what Maiman had done, Abella initially thought the Hughes report was a mistake. "It can't work in ruby, Schawlow told me it can't work," he said.

Sullivan explained that Maiman had used a flash tube, and Abella realized that Maiman had changed the rules. For an instant, the brilliant flash could do what Schawlow had said couldn't be done—depopulate the ground state. Abella admitted to Sullivan that it was possible. After he finished that call, Abella phoned Townes in Washington to say they'd been scooped. Then he and Cummins "shut everything down and went across to the West End and had more beers than I care to remember."

Bell Labs was astonished and initially skeptical. "It came as quite a shock when Maiman announced he had found lasing in ruby," Clogston says. Amnon Yariv, a member of the Bell laser group, was vacationing in San Diego at the time, so

Jim Gordon called and asked Yariv to drive to Malibu and check out the Hughes claim. Maiman seemed a bit nervous, but he nonetheless showed Yariv his data. Examining it carefully, Yariv saw that the fluorescence spot collapsed suddenly above threshold to a small laser spot, at the same time the brightness increased suddenly. He was convinced. "I called Jim back that evening to tell him that, in my opinion, Ted had made the world's first laser," he said later. It was a vital and deeply symbolic vindication.

Soon after the press conference, a reporter visited Bell to get their reactions to Maiman's announcement. Bell physicists talked him out of a copy of the preprint and studied it carefully. Some remained skeptical, but Schawlow—doubtless abashed that he had missed the possibility—was convinced Maiman had done it.

TRG was shocked; "nobody knew Maiman," says Steve Jacobs. Willis Lamb, who had settled into a prestigious professorship at Oxford University, heard only that a Californian had made a laser. His first thought was that it had been Wieder. Lamb hadn't heard from Maiman in a while, but Wieder had earlier mentioned his interest in ruby, and had recently written about his move to California.

Peter Sorokin heard the news on the radio. When he checked the details, "We were astounded at how he pumped it. You could put in some numbers and figure out he must have had megawatts of light. Here Stevenson and I were considering how to make the thing work on a threshold power of a few watts," said Sorokin.

The surprise was universal, and often accompanied by doubts. Within days after the press conference, Maiman was invited to describe the ruby laser to a joint meeting of the Institute for Defense Analyses and the Advanced Research Projects Agency. It wasn't the kind of invitation anyone working at a military contractor could afford to turn down, especially when he was trying to convince the world that he had delivered a breakthrough that the Pentagon had been seeking.

With Townes at IDA, Maiman got a different type of grilling than the press had given him. "Dr. Townes more or less conducted the questions and the presentation and was reluctant to believe that this was an operable laser," Maiman testified later in a patent dispute. Townes's skepticism was understandable. Until then, the only accounts he had seen of Maiman's ruby laser were in the general press. He and Schawlow had thought their analysis had covered the most important laser possibilities. Schawlow had assured him that ruby wouldn't work. Maiman's success with ruby had blindsided Townes, and that was a personal and professional embarrassment. But Maiman laid out the evidence and prevailed.

One reason for the surprise was that the east coast establishment knew little about Hughes Aircraft or Maiman, and less about his research. Their world was centered on the Northeast Corridor from Boston to Washington; California seemed far away in 1960. Like Schawlow, most establishment physicists missed the significance of Maiman's paper on ruby fluorescence in *Physical Review Letters*. (TRG was an exception, but they weren't really part of the establishment.)

A bigger surprise was that Maiman's laser produced short, very intense pulses. Among the other developers, only Gould had thought seriously about pulsed lasers, and they were only a small part of the TRG program. Schawlow had tried exciting red ruby with a flashlamp, but only because a continuous lamp would have overheated the crystal when it was cooled to liquid helium temperature—and in any case that experiment didn't work. Bruce McAvoy and Wieder had tried a similar experiment at Westinghouse but gave up after their lamp exploded.

Schawlow, Townes, Javan, and Gould all had proposed lasers that would operate continuously like light bulbs or radio oscillators, with a steady flow of input energy generating a steady beam of coherent light. Since population inversions seemed hard to produce, they assumed the gain from stimulated emission would be weak, and that the right conditions for laser oscillation would be hard to maintain. Except for Gould, they expected that a laser would emit only a modest amount of light. They had broken through the conceptual logjam of always assuming thermodynamic equilibrium, but they still assumed a laser would operate in a steady state. By scrapping that second assumption, Maiman made the breakthrough to the laser. In fact, Maiman had invented a whole new type of laser, which emitted coherent light in powerful bursts rather than a steady beam.

Pulsed operation opened new possibilities for both energy sources and laser materials. Energy sources like flashlamps can deliver much higher powers in a single burst than they can continuously. Likewise, transient population inversions are much easier to produce than steady inversions in many materials, particularly ruby. Pulsed operation allows time for the dissipation of heat, so Maiman didn't need to cool his little ruby laser. Yet only TRG had taken a serious look at pulsed lasers before Hughes held its press conference, and they hadn't done much with the idea.

Maiman's first laser beam didn't behave exactly as had been predicted. It formed a broad, horseshoe-shaped pattern on the screen, not a single, tightly focused spot. The beam didn't switch on sharply as the laser passed threshold. He explained those problems came from the poor quality of the ruby crystal. He supported his laser claims with his observations of line-narrowing and pulse-shortening that could only arise from stimulated emission. He may have left some doubts behind him in Washington, but he resolved them quickly, when he tested the new better-quality ruby crystals and saw the expected narrow beam and sharp laser threshold.

Townes later said he was surprised at how easy the ruby laser was to make. That annoyed Maiman, who knew it wasn't easy. Yet Townes's comment was also a back-handed tribute, because Maiman had made ruby *look* easy. He had found a light source well matched to his laser material. He had packaged them together to transfer energy efficiently between them. He didn't need all the cumbersome and complicated equipment that cluttered Javan and Bennett's helium–neon lab at Bell. Maiman's laser looked simple, but as a work of physics and of engineering, it was truly elegant.

The true test of any scientific research is whether anyone can duplicate it. Like the laser, cold fusion was announced at a press conference called by scientists who wanted to stake their claim to a new idea. Cold fusion quickly fell into disrepute when other scientists could not replicate the experiments. Maiman's design for the ruby laser was so simple and elegant that others quickly replicated it within weeks, as soon as they got their hands on the right components and adequate ruby crystals.

Nobody had blueprints. The groups trying to replicate Maiman's ruby laser initially had to rely on press reports and the Hughes photo showing Maiman's face behind the mid-sized flashlamp. The news stories claimed that was the flashlamp used in the first laser, so Bell Labs and TRG ordered the mid-sized lamps as soon as they identified them. Only later did the scientists get preprints of Maiman's paper.

Ron Martin and Dick Daly had a head start at TRG because they already had been considering ruby as an alternative to rare earths added to solid crystals when Maiman's paper on ruby fluorescence appeared in the June 1 *Physical Review Letters*. TRG was quick to realize its importance: bright pulses from a flashlamp could excite a significant fraction of chromium atoms from the ground state. Helical xenon flashlamps were fairly well known, so TRG got some and started banging away, but progress took time. Like Maiman, they had to contend with poor-quality ruby rods. "It was a brute force thing," recalls Martin. They were already testing the larger lamps when Hughes held the press conference. It took cooling the ruby rod with liquid nitrogen to put the TRG ruby laser over threshold a few weeks after the press conference. They needed to cool their ruby rod because the larger flashlamp produced much more heat than Maiman's. It was the first laser outside of Hughes, but TRG never got around to publishing a paper. They were working on a classified military contract, where priorities were different. They also knew Maiman had won the race and had his own paper in the works.

The initial atmosphere at Bell Labs was skeptical, with some scientists insisting ruby couldn't work. That began to change after Bell got a copy of the preprint of the *Journal of Applied Physics* paper from a magazine writer who visited the labs to interview people about their reactions to Maiman's laser. Yet the ultimate test was to see if they could duplicate Maiman's results.

Don Nelson wasn't sure what to think about ruby, but he wanted to try something else after his work on uranium crystals had stalled. He recognized the General Electric flashlamp in the Hughes publicity photos and immediately ordered one. Then he located a used power supply—which measured four by two by two feet and came on a wheeled cart—in New York and had it delivered to Bell in a week. The labs had plenty of pink ruby crystals, so he had one cut and polished to make a rod four centimeters long and a half-centimeter in diameter—longer and not as fat as Maiman's, more like a stubby pencil than a fingertip. Another Bell physicist, Robert J. Collins, soon joined him.

Like Maiman, they threaded their rod inside the coil of the lamp, and surrounded the assembly with an enclosure. Their more powerful lamp required cooling, so they surrounded it with a pair of concentric cylinders, and flowed water between them to remove the excess heat. They coated the inside of the inner cylinder with powdery white magnesium oxide, to reflect the light from the outside of the lamp coils. The white surface reflected the light diffusely, spreading it through the inside of the cylinder so more of it would hit the ruby rod.

Schawlow was aware of their work, and as the time neared for their first experiment, he got a call from TRG, telling him that Martin and Daly had succeeded in operating a ruby laser the night before. He went to tell Nelson and Collins, and they got talking about how to test for laser action. Nelson and Collins needed a high-resolution spectrometer to measure changes in the ruby spectrum above laser threshold. Schawlow had the instrument they needed, and agreed to help them with the measurements, although he still wasn't particularly interested in building a laser. Like TRG, they decided to cool the ruby rod in liquid nitrogen. The mid-sized flashlamp was big enough to hold a small transparent dewar filled with liquid nitrogen to cool their pencil-thin laser rod.

Their flashlamp was bigger than Maiman's, and it was rated to withstand a hefty 4000 volts, but that wasn't enough to reach laser threshold. The dewar got in the way, so less pump-light reached the rod than in Maiman's uncooled laser. The Bell team thought they were getting close, and cautiously cranked up the voltage further. At 4200 volts their instruments told them they had laser action. "It does not pay to be gentle when you have a threshold effect!" Schawlow observed, "You sometimes have to be fairly violent to get sufficiently far from equilibrium." Maiman's laser had crossed threshold at 950 volts applied across the flashlamp.

They succeeded without blowing anything up. But they also found something they didn't expect. While Nelson was away at his sister's wedding, Collins examined the structure of the laser pulse with an oscilloscope. Instead of a single pulse, he saw a series of ragged spikes, as if the laser was firing erratically. Puzzled, Collins wheeled the apparatus to Schawlow's lab so he could examine what wavelengths were emerging from the laser. They were still scratching their heads when Nelson returned and recognized the pulses as relaxation oscillations, optical aftershocks that followed the big pulse, an effect well known in microwave masers.

Many other details remained to be worked out. They had the sides of the rod finished roughly because Schawlow thought reflections from smoothly polished sides would break up the beam. They still couldn't see the beam initially, but that was because they had not enclosed the whole flashlamp, so the burst of white light washed out the red beam. Eventually they took a photograph through a red filter that showed a small, bright red spot a half of a degree wide that marked a beam, something which Maiman had not reported at the press conference. Nelson, Collins and Schawlow had to completely enclose the lamp and clean up the mirrors to produce the pencil-like beam we now expect from a laser. Soon they were firing the beam a quarter mile down hallways at night when they were alone in the sprawling Bell Labs building, sending the beam through a hollow pipe called a waveguide that Bell had built earlier to transmit high-frequency microwaves.

Meanwhile, Garrett and Kaiser set aside their original laser project to build their own ruby laser in a different Bell department some distance from Schawlow. It wasn't disorganization; it was a luxury the well-funded Bell Labs could afford on an important project. Their laser also produced a narrow beam, although they had polished the sides of the ruby rod, showing that Schawlow's rough finish wasn't necessary.

Others also tried to make their own ruby lasers. A week after the Hughes press conference, Townes asked Abella to build one to demonstrate to the Pentagon, while Cummins continued with cesium. Abella got a pink ruby sample from another Columbia physicist and a flashlamp from a photographic equipment store. Schawlow helped him organize the experiment, and in August he also had a working ruby laser.

Mirek Stevenson of IBM called Schawlow and a physicist he knew at TRG to pump them for information on ruby. Schawlow told him which flashlamp to order, and soon Stevenson and Sorokin had their own ruby laser running. It worked so well they decided to scrap the elaborate reflective cavity they had designed for crystal lasers and borrow the flashlamp approach. At the National Research Council of Canada, Boris Stoicheff stopped work on a mercury-vapor laser, and quickly made Canada's first ruby laser. At Harvard, Bloembergen asked his student, Robert A. Myers, to make a ruby laser, and Myers found "it was almost ridiculously easy to repeat" Maiman's experiment.

It didn't take long for ruby lasers to start scaling to serious power levels. The Boston-area aerospace company Raytheon set aside its work on rare earth crystal lasers when they heard of ruby. By good fortune, Raytheon had some big chunks of ruby available. They ordered a big flashlamp, and prepared a 12-inch-long ruby rod. Once they got the laser going, the results were impressive. Raytheon physicist Hermann Statz recalls pulses powerful enough to punch holes in a meat cleaver blade. Engineers were soon measuring pulse energy in "gillettes"—the number of razor blades they could penetrate.

While other groups played with their new ruby lasers, Collins, Nelson, and Schawlow decided to study theirs systematically and publish the results. They set themselves a deadline and began making measurements. Schawlow bought the best oscilloscope he could find and they used it to study how ruby emission dropped after the flashlamp pulse. They measured the irregular series of stuttering pulses that Nelson had noted earlier, and confirmed they were relaxation oscillations. They measured how narrow the range of wavelengths was in the laser pulse, and how coherent the light was across the end of the laser. They saw the fraction of emission on the R1 line rise sharply as the laser power increased. They tried operating the laser with the ruby rod at room temperature, and compared it to their measurements with the rod in liquid nitrogen. They also confirmed the beam was emerging from the laser in a narrow beam, only 0.3 to 1 degree across.

They hadn't finished all they wanted to do when the deadline arrived, but Nelson and Collins wrote up the results anyway, listing Schawlow as a coauthor. Garrett and Kaiser supplied some data, so laboratory etiquette dictated that they be added to the paper along with Walter Bond, who had been working with Garrett and Kaiser on other projects. Figure 17.1 shows Schawlow and Garrett with one of Bell's first lasers.

Obviously well aware of Maiman's problems in publishing his paper, and afraid they might suffer the same fate, the Bell group carefully avoided using the forbidden word "maser" anywhere in the entire paper so they could sneak it past Goudsmit. They titled it "Coherence, Narrowing, Directionality, and Relaxation Oscillations in the Light Emission from Ruby" to make it sound like the fundamental physics Goudsmit wanted to publish rather than descriptions of devices. Their strategy worked, although the Bell return address may have helped smooth the paper's way through the editorial mill. *Physical Review Letters* received it on August 26 and published it in the October 1 issue. The turnaround time was enviably quick, but Maiman's paper on ruby fluorescence had taken only a few days longer.

The paper relates some interesting observations from playing with a laser, but its viewpoint is centered on Bell Labs to the point of myopia. It opens by crediting Schawlow and Townes with proposing the laser concept, then says that Schawlow proposed "the use of a ruby rod for the observation of these effects." Yet the reference is to the Shawanga Lodge paper in which Schawlow said the pink ruby Maiman used wouldn't work! Nelson says the intent was to credit Schawlow for proposing the rod geometry, because at the time Maiman had only published that his ruby crystal was of "1-cm dimensions." Bell's rod was longer and thinner than Maiman's, and Schawlow thought those dimensions were vital to produce a laser beam. Nelson says that the fact that Schawlow had proposed using red ruby rather than the pink ruby used in the first laser was "of secondary importance at that point to us."

Maiman's existence is barely admitted. The Bell paper credits Maiman only with having "observed a decrease of the lifetime and a narrowing of the line shape for ruby fluorescence" and cites only the brief *Nature* report. Maiman thought those two observations proved that he had a laser. Bell claimed doubts based on Maiman's failure to report a sharp threshold, a narrow beam, or relaxation

FIGURE 17.1. Art Schawlow (left) and Geoffrey Garrett with Bell Labs ruby laser. (Courtesy of Bell Labs)

oscillations. Yet there was more to the picture by the time Bell submitted their paper in late August. Bell had obtained a preprint of Maiman's original paper weeks earlier. They should have mentioned it was in print at the *Journal of Applied Physics*, unless they had already seen it in the September issue of *British Communications and Electronics*. They also should have mentioned that Maiman saw a beam and distinctive threshold behavior as soon as he plugged one of the better ruby rods into his laser on July 20. Maiman says he told Collins about that experiment when Collins phoned him in late August to ask for the reference to the *Nature* letter.

Nothing in the published *Physical Review Letters* paper crosses the line to outright falsehood, but crediting Schawlow with proposing the ruby laser comes close. The laser Bell described was one that Schawlow had said could not work in the same paper Bell cited to say it was his idea. Schawlow had explicitly ruled out the chromium R lines in pink ruby because they dropped to the ground state. At Shawanga Lodge, he said the satellite lines in red ruby might work because they did not go to the ground state. Bell's laser worked on the R lines in ruby containing 0.05 percent chromium—pink ruby. Schawlow didn't deserve credit for that; Maiman did. The satellite lines Schawlow had proposed had not yet worked in a laser. Only by ignoring the critical distinction between red and pink ruby could Bell credit Schawlow for the ruby laser, yet Schawlow had stressed the importance of that distinction.

Bell instead focused on another distinction, the geometry of the ruby laser rod, which turned out to be based on a misinterpretation of Maiman's paper. Maiman didn't spell out the geometry of his ruby crystal—perhaps because he referenced Schawlow and Townes's original laser proposal. But it wasn't the short and stubby shape of Maiman's original ruby rod that limited its performance—it was optical flaws inside the crystal. Maiman confirmed that as soon as he got better ruby crystals, but Bell didn't pick that up when Collins talked with Maiman. Maiman's laser rod was like a short broken piece of a fat preschooler's crayon, while Bell's was the shape of a long thin crayon. But that distinction didn't matter in the end; both were working lasers.

Looking back, Nelson says, "The most important contribution of our paper was pointing out that the appearance of relaxation oscillations was the proof of laser oscillation," rather than mere amplification. In fact, Asawa and D'Haenens had seen the slowly decaying series of pulses shortly after the first experiments, but they didn't understand them, and Maiman initially thought they were an instrument problem. The next day they showed the pulses to Birnbaum, who recognized them as relaxation oscillations, but Maiman wanted Asawa and D'Haenens to explain the pulses theoretically before publishing their observations, and they never got around to it. Maiman did eventually report observing the pulses in the long analyses of ruby laser operation he published in 1961. Yet for all practical purposes that was merely the dotting of an "i" in the case for a working laser. Maiman's experiments have stood the test of time as proof of the first laser, and his tests of the second batch of ruby crystals nailed his case.

The real issue that was Bell scientists were in denial. They couldn't admit to themselves or to their management that Maiman had scooped them with a type of laser they had said wouldn't work. Bell Labs was the world's biggest and best industrial laboratory, but unlike the parent AT&T, the labs did not have a monopoly on their stock in trade—ideas.

That denial shows clearly in the paper. It fails to reference Maiman's unpublished paper, although Bell had a copy. It's possible Collins thought the whole thing was in *Nature*, which he had not seen, but Maiman probably told him otherwise. The normal protocol would be to list it as "in press" at the *Journal of Applied Physics*. It's unlikely Bell knew it was being published in *British Communications and Electronics*, although the September issue came out about August 15, while they were still preparing their paper. If Collins hadn't seen the August 6 issue of the widely circulated *Nature*, he was unlikely to have seen the comparatively obscure *British Communications and Electronics*. That wasn't the only omission. The Bell paper does not mention that Maiman had reported changes in the widths and intensities of the R_1 and R_2 lines before Bell did—a significant omission the referees should have caught. It also fails to mention that TRG had told Schawlow of their ruby laser before Bell's operated.

By sneaking the paper past Goudsmit and downplaying Maiman's work, Bell scientists might have been laying the groundwork for claiming they really had made the first laser. At least some of the group had not yet admitted to themselves that Maiman had gone all the way to the finish line of the laser race—full-blown laser oscillation. If they could somehow show that Maiman had not succeeded, they would be first in print with proof of a working laser oscillator.

There is no solid evidence that was Bell's intent, but it doesn't take a real conspiracy to raise suspicions. The slights and omissions were significant, and Maiman was primed to believe. He blamed Western Electric for stealing an invention his father had made while working at the Bell manufacturing subsidiary.

Proper publication of his full report on the first laser had been blocked by editorial bungling and bad luck. It's no wonder Maiman was suspicious.

Bell's public relations staff jumped into the action as soon as they heard Collins and Nelson had built their own ruby laser. Their job was to promote the glories of Bell Labs, and they had a good sense of what would capture the public eye. They convinced Collins and Nelson to haul their laser to the roof of an old radar tower at the Murray Hill, New Jersey, lab and send the beam to another Bell tower 25 miles away in Crawford Hill. Sending the beam through the air such a distance was a perfect example of how Bell hoped to use the laser for communications. It wasn't that simple in reality, and optical fibers of ultratransparent glass eventually proved a better medium for laser communications than air. But it was unquestionably a good stunt, and it diverted the public imagination from death rays.

A clear line of sight between the tower tops let Nelson and Collins aim the beam in daylight once they bought a little telescope to spot the Crawford Hill tower. Scientists at the other end could see the red flashes, and detect the beam, which had spread to 200 feet across over its 25-mile path. They needed a filter to block sunlight and a sensitive light detector to spot the red flashes, so it wasn't easy—but it worked. The demonstration of a potentially practical application also complemented their *Physical Review Letters* paper, which was full of details far more interesting to scientists than to the general public.

Bell headlined the communication demonstration in a release distributed at an October 5 press conference in New York, with the research paper relegated to page 3. The demonstrations got more ink, although press coverage varied widely. *Time*, which hadn't deigned to believe Maiman, devoted half of its Science section to Bell's "fantastic red spot," never mentioning that it wasn't the first. Conversely, the *New York Times* didn't mention Bell's 25-mile transmission until a week later, when it ran a longer report on Hughes's efforts to develop a laser radar system for the Air Force, which Maiman described at a Boston meeting on October 12. *Time* mentioned that high-power laser beams might nudge satellites between orbits, or be used to control chemical reactions, but said nothing about death rays. If anyone asked, Bell scientists would have downplayed the possibility.

As press releases go, Bell's announcement was even-handed. It credited Schawlow with proposing making lasers from ruby, but went on to say, "It was used in the manner originated by T. H. Maiman of Hughes Research Laboratories, who first observed optical maser effects in ruby." That went right by some reporters. *Time* didn't catch on and didn't mention Hughes or Maiman. Bell's preference for "optical masers" confused the usually knowledgeable *Electronics* magazine, which wrote, "The optical maser is similar in principle to the Hughes Aircraft laser." It was merely the Hughes laser with a Bell Labs label.

As the dust settled, Art Schawlow decided he owed it to himself to try making the red ruby laser he had abandoned after proposing it at Shawanga Lodge. He and George Devlin cooled a ruby rod with liquid nitrogen and illuminated it with a large flashlamp, producing laser pulses at 701.0 and 704.1 nanometers, the four-level laser lines that Schawlow had predicted should exist at low temperatures (fig. 17.2). By then it was mid-December, and it wasn't even the first four-level laser. Tired of trying to work his way around Goudsmit's ban on "maser" papers in *Physical Review Letters,* Schawlow sent his report to *Physical Review* in mid-December.

Yet even with the red ruby laser Bell couldn't claim an unqualified first. After frustrating months on the sidelines, Irwin Wieder had finally broken through the

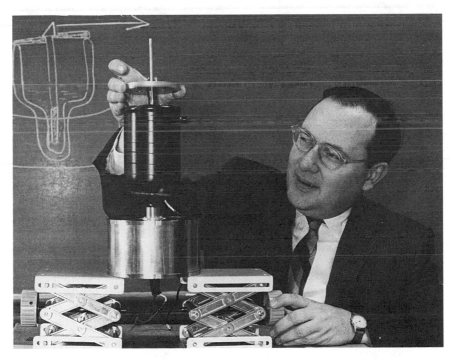

FIGURE 17.2. Art Schawlow demonstrates an early cryogenically cooled ruby laser; its structure is on the chalkboard in the background. (Courtesy of Stanford News Service Archives)

FIGURE 17.3. Irwin Wieder presents a paper on ruby physics at the second quantum electronics conference in 1961. (Courtesy of Irwin Wieder)

equipment freeze at Varian and obtained a sample of red ruby to build his own red ruby laser. It emitted at exactly the same wavelength as Schawlow's, and was also cooled with liquid nitrogen. Wieder (fig. 17.3) sent his report to *Physical Review Letters*, where it arrived December 19, the same day as Schawlow's. The editors decided they should treat the two the same, and ran them in the same issue of *Physical Review Letters*. However, by the time it got in print, other runners-up had begun to cross the laser finish line.

18

RUNNERS-UP CROSS THE FINISH LINE

RUBY PROVED A REMARKABLY USEFUL and versatile laser, and ruby lasers are still used today. However, no one thought ruby would be the ultimate laser. Maiman's success broke conceptual logjams and spurred the quest for the original goal of a laser capable of continuous operation. The runners-up were still pressing for the finish line.

Peter Sorokin and Mirek Stevenson at IBM were quick to order a flashlamp after they heard of Maiman's demonstration. After they built their own ruby laser, they turned to the brightly colored calcium fluoride crystals they hoped to make into lasers. Their spectroscopic studies had shown both were promising as four-level lasers, which they expected to work better than the three-level pink ruby laser.

The two boldly decided to abandon their elaborate square-resonator design. The powerful flashlamp should give them plenty of pump light to test how well the crystals worked in a laser. They had already shipped their specially grown crystals to an optics company that fabricated components from unusual materials. Now they phoned and told the company to cut and polish some of the raw crystals into rods rather than square reflective blocks. Then they asked the company to coat the ends of the rods with a layer of silver, to make mirrors for the laser light to bounce between. Custom optical fabrication is a delicate task, so it took a while.

When they got their rods back, they couldn't just slip them inside the coiled lamp and fire away. The crystals had to be cooled far below room temperature to distribute the uranium and samarium atoms properly among energy levels. That required putting the laser rod into an insulated cylinder filled with liquid nitrogen or another cryogenic fluid, with a transparent window that allowed light from outside to illuminate the cooled rod.

They first tested the rod tinted bright red by uranium atoms (fig. 18.1). They slipped it into the insulated cylinder and put the transparent window of the cylinder next to the flashlamp. It wasn't an efficient use of the lamp's energy. Most of the flash went in other directions, with only a small fraction focused onto the rod. Yet Sorokin and Stevenson expected their four-level laser system to make much more efficient use of the small amount of light that reached it. If they needed to focus more light onto their sample, they figured they could rearrange the experiment to concentrate more of the lamp's energy onto the rod.

Although the uranium atoms colored the calcium fluoride as red as ruby, the laser transition they wanted was invisible to the human eye, at 2.5 micrometers in the infrared. That made them even more dependent on instruments than laser experimenters using ruby, who were looking for red flashes. Sorokin and Stevenson aimed their rod at a detector which responded to infrared light, hooked it up to an oscilloscope, and watched the screen when they fired the lamp. The first time they tried, Sorokin saw the oscilloscope trace zoom off the scale as the infrared emission from the rod increased dramatically. The dark green calcium fluoride rod doped with samarium, which emitted at 0.708 micrometers, a wavelength faintly visible to the human eye, soon followed.

It was November, and the two young IBM physicists had reached an important milestone. They had made the second type of laser, and it differed critically from ruby. It was a true four-level laser, with the lower laser level separate from the ground state. Theory predicted that a population inversion should be easier to create in a four-level laser than in three-level ruby, and their experiment confirmed it was. The lasers had worked although most of the pump energy had gone elsewhere. They estimated their uranium laser needed only 1/500th as much input light to work as a ruby laser. With better optics and a brighter continuous lamp, a four-level laser should be able to emit a continuous beam. It was something to boast about.

Sorokin and Stevenson quickly wrote a paper on their uranium experiments, then pondered where to submit it to get swift attention. They were well aware of

FIGURE 18.1. Peter Sorokin (left) and Mirek Stevenson adjust the uranium-doped solid-state laser at IBM (Courtesy IBM)

Maiman's publishing fiasco and had seen Collins, Nelson, and Schawlow sneak their ruby laser report into *Physical Review Letters* by avoiding the word "maser." The mild-mannered Sorokin was not one to challenge anyone, but Stevenson was far more aggressive. When they finished their paper, Stevenson announced, "We're not going to send it. We're going to drive down to Brookhaven and tell Sam Goudsmit we want a decision before we leave." Sorokin tried to tell him that things weren't done that way, but in the end Stevenson persuaded him.

It was a decision well made. When they got to Brookhaven, they found Goudsmit "was slightly confused about the differences between masers and lasers," Sorokin recalls. Goudsmit was adamant that he didn't want another maser paper, but Stevenson was equally adamant that their paper was not about just another maser. Eventually Stevenson won, but Goudsmit didn't want the precedent to start a parade of eager scientists personally delivering their latest and greatest achievements. As the two prepared to leave, Goudsmit told them, "Next time, tell your people from IBM not to come down here with machine guns."

Stevenson evidently lit a fire under Goudsmit. The paper he delivered on November 28 appeared less than three weeks later in the December 15 issue of *Physical Review Letters*. The usual turn-around was four to six weeks. IBM heralded the news with a December 14 press conference at New York's Gotham Hotel, and the story made the inside pages of the *New York Times*. The IBM press conference came a day after Javan, Bennett, and Herriott finally got the helium–neon laser to work. That achievement had been a long time coming.

Maiman's demonstration of the ruby laser had been a disappointment, but Javan, Bennett, and Herriott could take solace in the fact that they had a different goal, a laser generating a continuous beam. Ruby couldn't do that. Although the ruby laser did produce nominally coherent light, pulsed operation limited its coherence. The longer a laser operated, the better stimulated emission lined up its light waves so they all marched exactly in phase. The helium–neon laser promised more coherent output, and looked much better for communications, which was the basic business of Bell Labs. However, getting it to work was a tough problem.

Two crucial steps had been achieved by July. Bennett and Javan had measured laser gain in a mixture of helium and neon, showing that, in principle, stimulated emission should build up as light bounced back and forth between mirrors. That sign of progress reassured some budget-conscious managers, although Javan remembers some doubters who didn't believe their measurements. Meanwhile, their optical wizard, Don Herriott, had designed mirrors able to reflect 99 percent of the light that struck them. Javan and Bennett needed those nearly perfect mirrors to push their laser across the threshold for oscillation.

Making the mirrors required depositing 13 very thin layers on pure silica plates polished so smoothly that bumps rose only 1/100th of the wavelength of light above the surface. Frances Turner made them at Bausch & Lomb by depositing a layer of magnesium fluoride a fraction of a wavelength thick, then an equally thick layer of zinc sulfide, and repeating the process until 13 layers were deposited. That structure made the mirrors reflect very strongly at a limited range of wavelengths, at 1.1 to 1.2 micrometers where they expected neon atoms to produce stimulated emission. Other optics researchers had been developing such multilayer mirrors for years, but the number of layers Herriott wanted pushed the state of the art, and the mirrors proved difficult to make. Mirrors containing so many layers tended to form craze patterns, spider webs of indentations like the alligator-skin texture of surfaces covered with too many layers of old paint. Those flaws posed a serious problem.

Several important tasks remained before they could get the laser working. The gases had to be purified to exacting standards. A structure had to be built, with a glass tube to contain the gas, mounts to hold the mirrors, and a way to excite the gas. The tube, mirrors, and laser gases had to be assembled into a perfect cavity without damaging the delicate reflective coatings. Air had to be kept out of the tube containing the laser mixture. And everything had to be tweaked and fine tuned until it was adjusted to perfection. As they closed in on their target in the summer, Herriott put aside his other work and started working full time on the laser project.

As the optics specialist, Herriott faced some daunting problems. Gain in the mixture of helium and neon was low, so the tube had to be about a meter long in order to get enough gain to reach the oscillation threshold. But the mirrors were only 1.5 centimeters in diameter, matching the inner diameter of the quartz tubing that held the laser gas. The good news was that Fox and Li had shown that little light would leak out the sides if flat mirrors were perfectly parallel to each other, and perfectly perpendicular to the path of light through the tube. The bad news was the word "perfectly." If the two mirrors were out of alignment by just a few seconds of arc—the size of a dime seen from a mile away, or the planet Neptune viewed from the Earth—the loss would become too large for the laser to oscillate. Today it's possible to align mirrors to that exacting precision, but it requires a working laser that generates a steady, coherent beam. Herriott didn't have one, so he had to mount the mirrors so they could be tilted back and forth until they came into alignment. Maiman had not suffered the same problem because ruby has a much higher gain than the mixture of helium and neon, so perfect mirror alignment wasn't critical.

To meet those demanding requirements, Herriott, Bennett, and Javan designed a 60-centimeter quartz tube with mirrors mounted in stainless steel metal bellows on each end. Precision micrometers built into the mirror mounts let them adjust the tilts of the mirrors. It all had to be arranged very carefully, so a relatively large motion of the micrometer moved the mirror just a tiny bit. The whole assembly had to be able to withstand being baked at high temperatures in high vacuum to drive contaminants from the inner wall of the tube and the metal housing. And it had to be sealed tightly so air couldn't seep inside.

A tight seal would have been difficult to make if they had to deliver electric current into the gas through wires. Fortunately, they had another technique that didn't require wires. Radio waves and microwaves penetrate glass, and are absorbed by electrons in helium atoms. Turn up the intensity, and the electrons absorb enough energy to escape from the helium atoms, and those electrons can excite other helium atoms which then transfer energy to neon, producing a population inversion.

In their gain experiments, Bennett and Javan had used a 10-megahertz radio oscillator, but Javan thought microwaves would work better, and bought a microwave tube generating frequencies in the 2- to 4-gigahertz range. The impetus may have come from another round of budgetary restlessness in upper management, disconcerted that helium–neon seemed so hard while ruby seemed to work so easily. Al Clogston rescued the helium–neon project, but Bennett and Javan could hear the clock ticking. The microwave generator arrived in August, and they set it up to excite the gas inside their laser tube. When Javan pulled the switch, the tube absorbed so much microwave energy that it melted. It was back to the drawing board—and to the radio-frequency source.

Meanwhile, measurements of optical and gas properties continued. In early summer Javan and Bennett had hired a new lab technician, Ed Ballik, who had dropped out of high school and worked for eight years before going to college and earning a degree in engineering physics. The three pulled long hours, closing down the lab and walking through deserted hallways to the cafeteria for a snack before going home at 11 P.M. They knew they had competition. In the fall, physicists at the Lebedev Physics Institute in Moscow claimed they had observed stimulated emission from a mixture of mercury and zinc vapor. The Russians turned out to be wrong, but nobody knew that at the time.

It took weeks to put together another set-up. Herriott went through the stock of mirrors from Bausch & Lomb, picking the two best to put into the ends of the tube. When everything was in place, he put it into an oven in high vacuum to bake out impurities overnight. The next day, after the tube cooled, he examined the mirrors, and found that the fragile coatings had cracked and crumbled, leaving bare glass in place of the reflective layer.

The mirrors were the most critical components, so this failure was as much a disaster as melting the glass tube. They changed their plans for baking out impurities, installed their next-best set of mirrors in an 80-centimeter tube, and tried again at lower temperatures. This time the heating and cooling left a pattern of lines on the mirror surfaces, giving them a crazed look, but they were mechanically intact. That was discouraging, but the tube was ready; they decided to go ahead and try it anyway. It was Tuesday, December 13, and with the holidays coming it would be well into the new year before they could get another experiment ready. Bennett, Javan, and Herriott thought they were on the right path, but they couldn't be sure how long management would continue pouring money into the expensive project. They filled the tube with helium and neon and set up their equipment.

They couldn't look for a beam, because the 1.15-micrometer neon transition they sought is in the invisible infrared part of the spectrum. They aimed the tube output into a monochromator, which was hooked up to a detector that would display the signal on an oscilloscope. By adjusting the monochromator, they could scan across the spectrum of light the tube was supposed to emit, watching the oscilloscope scan for results. But the first step was to look for output as they tried to align the mirrors.

Their careful gain measurements had yielded a wealth of information on energy transfer within the mixture of helium and neon, and the lifetimes of the important energy states. That had yielded a curious result with practical consequences—the gain was a few times higher if the laser was switched on and off hundreds of times a second than if the laser was operated continuously. The difference was hard to measure because they were switching the gain off and on, but it was significant.

That increase was important. Their early measurements showed that it would be tough to make gain of a continuously excited helium–neon mixture high enough to offset the losses of the optical cavity. Switching the radio-frequency excitation on and off quickly raised the gain enough to make prospects for laser oscillation look much better. For a while they worried that helium–neon mixtures might work only if they were pulsed. By December they were more optimistic about a continuous laser, but pulsed operation was still useful. The higher gain gave a stronger signal to help them align the mirrors, so they could pick up an output when the mirrors weren't as perfectly aligned as they had to be for continuous laser operation. Then they could adjust the mirrors to get the highest possible pulsed power—and then switch to continuous operation to see if it worked.

As the radio-frequency oscillator switched on and off, they adjusted the mirrors and the monochromator, but the oscilloscope showed no sign of laser action. They adjusted the mirrors and tried again, but again found no sign of laser action. They continued, adjusting the mirrors, scanning the monochromator, and seeing no sign of laser action on the oscilloscope.

By 4 P.M., Bennett, Herriott, and Javan were getting discouraged and stopped to discuss what to do next. Heavy snow was falling outside and it was getting dark, making it tempting to call it a day and head home. It looked like they would have to write off the flawed mirrors and assemble a new laser tube with fresh mirrors. Herriott fiddled with one of the mirror adjustments to keep his hands busy as they talked. Suddenly, Javan said "My God, what is that?"

Javan had glanced at the oscilloscope screen and seen a new type of signal appear, like the one they had hoped the laser would produce. "Our immediate response was to not touch anything and decide what to do," Herriott recalls. They had done something right, but they didn't know quite what, and after a long and frustrating day of getting nowhere, they didn't want to lose the moment of success. They plotted their course carefully before doing anything more.

Their first step was to adjust the monochromator to identify what wavelength they were seeing. It was 1.153 micrometers, one they had predicted could oscillate. Very gently, they adjusted their mirrors, carefully watching their oscilloscope display, to find where they got the peak signal. Then they scanned the monochromator and found that the laser was oscillating at four other predicted transitions. With the mirrors aligned, they switched to continuous operation and the laser oscillated, even with the lower gain.

Word spread quickly through the lab, and people began stopping in to see what had happened. The visitors included top Bell Labs brass, some of whom had been growing skeptical that Javan, Bennett, and Herriott would ever succeed.

They toasted their success with a bottle of 100-year-old Madeira port wine. Ballik had bought it shortly after joining the laser project, and stashed it in a filing cabinet in anticipation the laser would work. Javan invited the head of Bell Labs to share in the toast, not realizing that management had banned alcohol from the premises a few months earlier. A hasty memo exempted 100-year-old wine from the ban, and the toast went on. The wine was a disappointment, but the real reward was the triumph of making the helium–neon laser operate after all their long hours of hard work (fig. 18.2).

The next day, they started experimenting with the first continuous laser and found that one of the five wavelengths it emitted was considerably stronger than the rest. The strongest line was the easiest to produce, but it still took some effort to get the mirrors aligned precisely enough for the laser to work well. The eventually found a viable technique. "We simply banged (gently) on the laser after getting it nearly aligned while looking at the strong 1.15-micrometer line," Bennett recalled. As the tube vibrated, the mirrors moved in and out of alignment, and the laser flickered on and off. By carefully adjusting the micrometer screws

FIGURE 18.2. Don Herriott adjust optics of the first helium-neon laser with a beaker of celebratory liquid in his other hand, as Ali Javan and Bill Bennett look on. (Courtesy William R. Bennett Jr.)

as the laser flickered, Bennett and Herriott fine-tuned the laser until it generated a steady beam.

They also discovered their laser was so sensitive to vibration that noise near the mirrors would shift them in and out of alignment, varying the beam intensity. Working at the world's largest telephone company, the three were quick to realize they could use the effect to transmit voices. If they spoke near the sensitive mirror, the sound waves vibrated the mirror, causing the beam's intensity to vary along with the sound of the speaker's voice. Bennett and Herriott hooked the output of their light-sensitive tube to a phone line and spoke next to the mirror to Javan on another phone. They knew they should echo Alexander Graham Bell's first words over the telephone, but they weren't sure they remembered the right ones. "The first thing we said over this optical beam was, 'Come here, Watson, I need you,' and every other permutation of these and similar words to be sure we included the right phrase," recalled Herriott.

They spent the days before Christmas playing with their new toy, then submitted a report to *Physical Review Letters*. With Stevenson having beaten

Goudsmit into submission, Bennett and Javan felt free to put "optical maser" into their title, which stressed their most important first, that their laser had oscillated continuously.

Bell duly staged a press conference in Manhattan on January 31, 1961, the day before the paper came out. Bennett and Herriott brought the laser and got it running. They and Javan described how it worked, and Charles Townes came to endorse the new device, which he and Bell Labs insisted should be called an optical maser. They carefully explained the importance of a continuous beam that could be modulated to carry voices or other signals. Only Townes and Javan's names made the 15-inch story in the *New York Times* headlined "Bell Shows Beam of 'Talking' Light." But the report of the first continuous laser was buried on page 39, next to a report on a baby walrus named Ookie that had her tusk X-rayed. NASA had claimed the front page with a bit of monkey business, by sending a chimpanzee named Ham on a 16-minute flight into space in the second Mercury capsule.

The laser race was fading as news. Bell Labs, once the favorite, had finished in fourth place, out of the money in any horse race, behind Maiman and behind Sorokin and Stevenson's two lasers. Yet Bell did deserve a share of the glory. Javan, Bennett, and Herriott had the first laser that could generate a steady beam of highly coherent light, the original type that Gould, Schawlow, and Townes had envisioned. It also was the first laser to operate in a gas, and the first to be powered by electricity rather than by light from an external source. In the long run the helium–neon laser would far outclass the two IBM lasers, which never found any practical uses.

Nearly a year would pass before anyone outside of Bell succeeding in making their own working helium–neon laser. It took extraordinary experimental skill to overcome the low gain of the helium–neon gas mixture to push stimulated emission above the laser threshold.

Other developments helped reduce the barriers. While Javan, Bennett, and Herriott were struggling with the alignment of a pair of flat mirrors, other Bell Labs scientists were finding a way around the problem. Gary Boyd and his boss Jim Gordon, who had built the first microwave maser with Townes, did a detailed mathematical analysis of how light would bounce back and forth between a pair of curved mirrors arranged so each focused light onto the surface of the other. French astronomer Pierre Connes had thought the curved mirrors had to be placed within 0.01 millimeter of the ideal spacing, and TRG had accepted him at his word. But

Boyd and Gordon showed the tolerances were much larger. Curved mirrors vastly simplified laser design, and they quickly became standard on gas lasers.

TRG was falling behind. While Maiman, Sorokin, Javan, and Bennett were building lasers, Gordon Gould was stuck on the sidelines, waiting for word on his clearance from the bureaucratic morass of the Pentagon. Months dragged on, and troubles grew in Gould's marriage. Locked out of TRG's laser labs, Gould told his troubles to an attractive woman whose desk was near his. Marcie Weiss sympathized with him, and the two plunged into an affair. The fact that Weiss was TRG's corporate security officer made that a very dangerous affair, but that didn't deter Gould. His friends recall that the security mess seemed to muddle Gould's mind and cloud his technical judgement; perhaps it affected his romantic life as well. Or perhaps the part of Gould that liked to flout convention enjoyed the multiple levels of illicitness of the affair.

Jacobs and Rabinowitz struggled on with cesium, getting advice and ideas from Gould although he couldn't visit their lab. Around the time that helium–neon worked at Bell, they detected convincing evidence of a population inversion at 3.2 and 7.2 micrometers in cesium sealed in a Pyrex cell, which they first reported in March 1961 at an Optical Society of America meeting in Pittsburgh. Cummins thought he saw hints of a population inversion in cesium, but his evidence was inconclusive.

As Daly and Martin turned from developing new lasers to studying the properties of ruby lasers, Jacobs and Rabinowitz plodded onward, trying to demonstrate amplification in cesium vapor. They expected much more gain at 7.2 micrometers, but the available detectors couldn't measure the increase in power at that wavelength. Stimulated emission was weaker at 3.2 micrometers, but the detectors were better. They tried to make an oscillator, but they couldn't align their flat mirrors precisely enough. They tried exciting the cesium with pulsed lamps, but finally returned to their continuously operating ultraviolet helium lamp. After finally observing amplification in the summer of 1961, Jacobs and Rabinowitz hand-delivered their report to *Physical Review Letters*, but had to haggle with the referees to get it published. Security restrictions began to ease with the election of John F. Kennedy. Gould's clearance languished in the bureaucratic morass even after Yarmolinsky was named a special assistant to Defense Secretary Robert S. McNamara. But with other lasers reported openly, the Pentagon finally declassified the cesium laser, and Gould at last could work on the project.

By then, other runners in the laser race were reaching the finish line. Eli Snitzer demonstrated a laser made from a thin rod of glass doped with a dash of the rare earth neodymium; pumped by a flashlamp, it fired pulses like Maiman's ruby laser. So did a crystalline laser made of calcium tungstate doped with neodymium, demonstrated Leo F. Johnson and Kurt Nassau at Bell Labs.

Cesium finally lased in the wee hours of a Saturday morning in March 1962. Rabinowitz and Jacobs got a big boost from Boyd and Gordon's study showing that curved mirrors could be aligned far more simply than had been thought; they had enough trouble with the laser medium without having to struggle with the optics. They tried for laser action on the seven-micrometer line, where they expected the threshold would be lower. They set up the experiment late Friday night. The oscilloscope trace told the tale well after midnight, jumping when they fired up the lamp. They continued collecting data, and tried to call Gould, but he was away for the weekend. On Monday morning Jacobs and Rabinowitz arrived at Gould's office, laboratory notebook in hand, to show him the data. He could read the results on their faces, and beamed a smile before they could say anything. "Well, I'll be damned. You made it work!" (fig. 18.3).

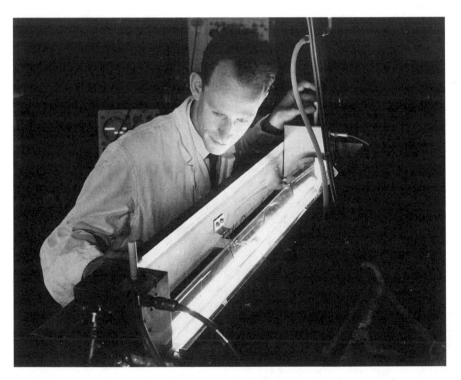

FIGURE 18.3. Steve Jacobs with TRG's cesium laser (Courtesy of Steve Jacobs)

19

EPILOGUE

THE YEARS THAT FOLLOWED would bring many more types of lasers, but ruby and helium–neon soon became the mainstays of laser research and applications. The two nicely complemented each other. Maiman's elegant design made ruby simple and robust; its short, powerful pulses were valuable in research and industry. Bennett, Javan, and Herriott's helium–neon laser was also a winner, generating a few milliwatts of light in a continuous and highly-coherent beam that was useful in a variety of research situations and alluring for communications. Helium–neon lasers took off rapidly after two other Bell physicists, Alan D. White and J. Dane Rigden, in 1962, made a version that emitted a visible red beam. Red helium–neon lasers became the standard for laboratory demonstrations and were the first type mass-produced for use in supermarket checkout scanners. Both helium–neon and ruby lasers have passed their prime after more than forty years, but remain in use.

The cesium-vapor laser served only as a proof of principle. Like the cryogenical-ly cooled red ruby, samarium and uranium lasers, it became a historical footnote because its operation was far too cumbersome to be practical. Lasers based on neodymium eventually caught on, but not in the same forms that Snitzer and Bell Labs demonstrated. In the fall of 1962, four groups raced to make the first semicon-ductor lasers in the shadow of the Cuban missile crisis. The first could fire only brief pulses and had to be cooled with liquid nitrogen or they would self-destruct. Today, they operate happily at room temperature inside compact disc (CD) players, in pen-sized laser pointers, and in supermarket checkout systems. Other types of lasers also emerged, some of them exotic. Peter Sorokin showed that dye molecules, dissolved in organic solvents, could work as lasers. Art Schawlow in a playful mood put the dye molecules in gelatin to make what he called a Jell-O laser, but also made serious use of dye lasers as exquisitely sensitive spectroscopic probes of processes inside atoms.

The microwave maser did not fare as well as the optical laser. In 1962, Charlie Asawa, Irnee D'Haenens, and Don Devor at Hughes reached Maiman's original goal, a microwave maser excited by light, by pumping ruby with a laser (figs. 19.1, 19.2). Today microwave masers are used in atomic clocks, and to amplify faint microwave signals from spacecraft exploring other planets. But they have few other applications and have slipped out of the public view.

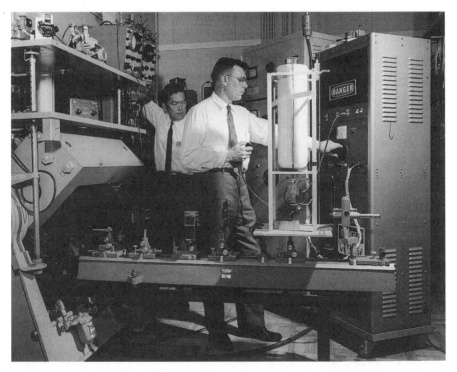

FIGURE 19.1. Charles Asawa (left) and Donald Devor ready a test of the Hughes laser-pumped microwave maser. The ruby laser is below the white cylinder of liquid helium at Devor's right. Light went across the optical bench in the foreground and onto a ruby microwave maser contained in a magnet. Asawa is adjusting the magnet. (Publicity photo courtesy of Hughes Research Laboratories/HRL Laboratories LLC)

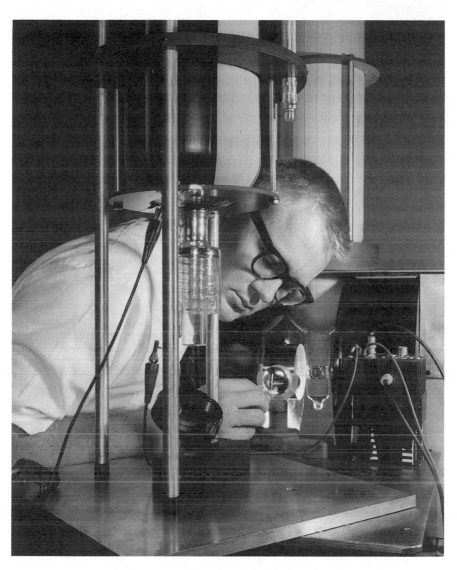

FIGURE 19.2. Irnee D'Haenens adjust the optics that direct a laser bean that pumps a microwave maser at Hughes Research Labs. (Publicity photo courtesy of Hughes Research Laboratories/HRL Laboratories LLC)

The optical laser, in contrast, caught the public imagination quickly and has remained highly visible ever since. It was all too easy to visualize the laser as a ray gun from pulp science fiction or B movies, especially when the popular press published wildly speculative articles about incredibly powerful lasers. Laser physicists grew annoyed with the need to continually debunk the myth of laser death rays. Art Schawlow, as was his wont, made a joke out of it. After a Sunday newspaper supplement published a wildly speculative article titled "The Incredible Laser," he taped a copy on his laboratory door, and pasted the words "for credible lasers, see inside" beside it, shown in figure 19.3.

The article reflected the times, an era of sometimes euphoric technological optimism, tempered only by occasional warnings of nuclear doom. The space race fuelled it, powered by John F. Kennedy's promise to put men on the moon, and the awe of seeing science fiction turned into fact. Exaggeration followed. At a time when laser pulses could punch holes in razor blades, the General Motors Futurama exhibit at the 1964 World's Fair in New York visualized a future world where a pair of high-energy lasers cut giant trees at their bases to make way for a huge road-building machine that paved a superhighway through a rain forest in a single pass.

Inevitably, laser reality fell short of the hype. Irnee D'Haenens's joke that the laser was "a solution looking for a problem" resonated through the little world of laser developers, who chuckled as they passed it along to their colleagues. Soon everyone in the growing laser community knew the line, but its origin was lost. Like the best wisecracks, it spread because it had a certain telling truth at its core. The laser was a wonderful and elegant device, but it wasn't developed to fill any specific need.

The first lasers were basically laboratory toys. Physicists played with them to study the properties of light and matter. Engineers played with them to see what they could do with focused coherent light. They learned interesting things. Peter Franken focused a red ruby beam onto a piece of quartz and the quartz turned a little of the red light into ultraviolet, a process called harmonic generation. Harmonics are familiar in sound, but no one had seen optical harmonics before. Emmett Leith and Juris Upatnieks at the University of Michigan's Willow Run Laboratory used the coherence of helium–neon laser to create a three-dimensional image called a hologram that has fascinated people ever since. Eye surgeons found that a laser pulse could reach inside the eye without cutting it open to obliterate the mesh of abnormal blood vessels that obstructs the vision of diabetics. Laser beams drilled holes in industrial diamonds and baby-bottle nipples, materials too hard or too soft to drill mechanically.

The Incredible Laser

FOR CREDIBLE
LASERS SEE
INSIDE

FIGURE 19.3. Art Schawlow posted this response to wildly exaggerated reports of prospects for laser weapons on his laboratory door at Stanford (Courtesy of estate of Arthur L. Schawlow)

Not everything worked, of course. The laser might be the greatest thing since sliced bread, but when engineers tried slicing bread with a laser, they got burnt toast. The bright promise of optical computing has been just around the corner for 40 years. We don't have laser saws in our workshops, and our police officers don't have laser guns in their holsters. But we can't expect lasers to do everything.

Along with the transistor and the satellite, the laser became a symbol of the technological revolution that has transformed life in developed countries since World War II. It made the reputations and careers of the men who helped create it.

Ted Maiman stayed at Hughes only months after his laser success. In early 1961, he jumped at the chance to become vice president of the applied physics laboratory at a new company in Santa Monica called Quantatron. Seven Hughes colleagues joined him in an effort to produce commercial ruby rods and lasers. Maiman's laser group grew to 35 people, and when Quantatron ran into trouble the next year, he got backing from Union Carbide to spin off the laser group as a separate company called Korad, with Maiman as president. It was one of the first commercial laser companies. After several years, Maiman left Korad to become a consultant. Later, he spent several years as vice president of advanced technology at the aerospace company TRW before returning to consulting. He tells his first-hand story of the laser race in his interesting autobiography, *The Laser Odyssey*.

D'Haenens (fig. 19.4) returned to graduate school at Notre Dame in 1963, courtesy of a Hughes fellowship that his laser work helped him earn, but didn't

FIGURE 19.4. Charles Asawa (left) and Irnee D'Haenens looking back at the first laser, circa 1985. (Publicity photo courtesy Hughes Research Laboratory/HRL Laboratories LLC)

specialize in lasers because he figured he had been through the exciting stage already. He returned to Hughes in 1966 and remained there until he retired to live in a scenic and exclusive trailer park overlooking the Pacific in Malibu. Asawa (fig. 19.4) stayed at Hughes after finishing his doctorate at UCLA, then moved to TRW in 1980, where he became involved in fiber-optic research. He retired from TRW, but remains active in fiber optics.

While still at the Institute for Defense Analyses, Charles Townes coauthored a paper suggesting interstellar communications via "optical masers," symbolizing his long-standing interests in laser and maser physics and infrared and radio astronomy. When his two-year term at IDA ended in September 1961, he was named provost of the Massachusetts Institute of Technology. The job put him at the top of the list of candidates to be the next president of MIT, but when the time came he was not offered the job. The University of California at Berkeley then lured him to join its physics faculty, where he remains active, a man of amazing energy, well into his 80s (fig. 19.5). He has served on countless government advisory boards, collected a long list of honors, and some of his students have become top figures in the laser field. His name is in the news even as I finish this book, joining the board of trustees of the SETI Institute and advising NASA on the future of the Hubble Space Telescope. He recounts his busy life in research and as a government advisor in an autobiography, *How the Laser Happened: Adventures of a Scientist*.

FIGURE 19.5. Charles Townes as an elder statesman of science and an active infrared astronomer, in his Berkeley office in 1988. (Courtesy Bell Labs).

FIGURE 19.6. Art Schawlow as an elder statesman of science, playing with a laser in his Stanford lab. (Courtesy Bell Labs)

Art Schawlow left Bell Labs to become a full professor at Stanford in 1961 after receiving offers from eight universities. He had no desire to be a maker and shaker like his brother-in-law Townes. His heart was in spectroscopy and academic research, and the San Francisco Bay area offered the best treatment he could find for his autistic son. His playful demonstrations and quick wit made him a popular professor (fig. 19.6), and his cutting-edge research in laser spectroscopy attracted stellar graduate students. He was a professor emeritus when he died April 28, 1999.

Ali Javan and Bill Bennett went their separate ways after developing the helium–neon laser, both leaving Bell Labs for professorships. Townes lured Javan to the MIT physics department, where he was active in basic research for many years; he is now a professor emeritus. Bennett returned to Yale, where his research spanned engineering and physics; he is also a professor emeritus.

Peter Sorokin (fig. 19.7) spent his entire career at the IBM Watson Research Center. The company named him an IBM Fellow, its highest honor for a scientist, allowing him to pick his own research directions. He is now retired. Mirek Stevenson left IBM in the early 1960s when the company warned him that the mutual fund he was managing on the side was a conflict of interest. He spent the rest of his career in financial services until his death around 2001.

FIGURE 19.7. Peter Sorokin at IBM circa 1985. (Courtesy IBM)

Nicolaas Bloembergen (fig. 19.8) turned his attention to spectroscopy and the interaction of light with matter. His keen analytical skills helped him develop elegant explanations for the complex nonlinear interactions evident between laser beams and materials. He remained at Harvard until he retired at age 70 in 1990, then moved to the University of Arizona, joining Willis Lamb who had also settled there.

Alexander Prokhorov and Nikolai Basov became leading figures in the Soviet science establishment and remained prominent in Russian science until their deaths. Prokhorov became head of the General Physics Institute in Moscow, while Basov became head of the Lebedev Physics Institute. Although on the same site, the two institutes were separate centers under the Soviet Academy of Sciences and the Russian Academy that succeeded it. Both were Communist Party members and full academicians, the highest rank for scientists in their system. Prokhorov was revered by younger scientists for maintaining his personal integrity within the Soviet system. He turned down a nomination to the Soviet Parliament. In a tape recorded to mark his 85th birthday, he recalled, "I said I do not want to be a deputy, because I am not a politician, I am a scientist." He died January 8, 2002. Basov did join the parliament, and later became a member of the Supreme Soviet, where he promoted science and education. He died July 1, 2001.

Valentin Fabrikant had little contact with the West throughout his lifetime, although he was well-known in the Russian physics community. Living through

FIGURE 19.8. Nicolaas Bloembergen celebrates receiving the 1981 Nobel Prize in Physics. (Courtesy AIP Emilio Segre Visual Archives, Bloembergen collection)

the purges of the 1930s left him a cautious man, and at least one observer said the Soviet establishment discouraged Westerners from contacting him. Fabrikant spent his career on the physics faculty at the Moscow Power Engineering Institute, where he was known more for his teaching than for his research. He was a full member of the Russian Academy of Pedagogical Sciences, but was never elected to the select Academy of Sciences, which did give him an award for his gas discharge research. Prominent Russian physicists studied under him and regarded him highly. In his later years, he wrote many popular science articles; he died March 3, 1991, at age 83.

Gordon Gould never received a security clearance. TRG gave up in August 1962, after an FBI agent told its attorneys that clearing Gould "would not be in the national interest." By then no one was surprised at the conclusion. The damage to TRG's laser program and Gould's career had already been done. Had Schawlow and Townes not opened the laser race by publishing their proposal openly, denying Gould clearance to work on the project he conceived might have delayed development of the laser for years. (TRG's cesium laser depended on Bell's curved mirrors.)

The Pentagon finally eased its classification of TRG's laser research about the time the FBI closed the books on Gould's clearance application. They hadn't done a good job of keeping secrets safe, Gould found when he asked them to return the original notebooks confiscated in 1959. "I discovered they'd lost one. How do you like that? It's supposed to be high-level security, and one of those notebooks was gone," Gould recalled. Fortunately, he had seen the problem coming and surreptitiously made copies for himself, and he still had those—without the CLASSIFIED stamps. He stayed at TRG for a few more years, then spent a few years on the faculty of Brooklyn College before accepting a buyout and joining with Bill Culver to form a small company called Optelecom in Maryland to manufacture fiber-optic equipment.

Gould's quest for laser patents spanned decades. He received his first patent on the laser in Belgium in 1962 and a series of British patents in 1964. But his critical U.S. patent was blocked by a series of "interferences" with other patent applications. Only after a lengthy series of legal maneuvers—detailed nicely in Nick Taylor's *Laser: The Inventor, The Nobel Laureate, and the Thirty-Year Patent War*—did Gould (fig. 19.9) prevail with a series of four patents covering most types of lasers and many laser applications. He sold off much of his interest to pay for the litigation, but eventually his earnings probably ran into tens of millions of dollars. Ironically, part of Gould's proceeds came from the laser company Dick Daly had founded after he left TRG, Quantronix.

FIGURE 19.9. Gordon Gould after receiving his laser patents in the 1980s. (Courtesy Gordon Gould)

FIGURE 19.10. Ted Maiman receives the Fannie and John
Hertz Science Award from President Lyndon Johnson in
1966. Ali Javan, at right, shared the award. (Courtesy Fannie
and John Hertz Foundation)

Many honors followed as well. Gould, Maiman, Schawlow, and Townes all
have been inducted into the National Inventors Hall of Fame. Maiman and Javan
received the Fannie and John Hertz Science Award from President Lyndon Johnson in 1966 (fig. 19.10). Maiman won the Japan Prize in 1987. The list of honors
could go on for pages.

The Nobel Prize in Physics came in 1964, a short time by Nobel standards,
just four years after the laser and a decade after the microwave maser. A rumor
swept the Stanford physics department a few days before the announcement that
the prize would go to Townes, Schawlow, and Maiman. Many were amazed when
it went instead to Townes, Basov, and Prokhorov "for fundamental work in the
field of quantum electronics, which has led to the construction of oscillators and
amplifiers based on the maser-laser principle." Frances Townes, more outspoken

than her reserved husband, declared "we were robbed," probably referring to Schawlow's exclusion. Maiman clearly feels he was robbed.

The Nobel archives remain sealed, but the phrasing and the names of the laureates give insight into the reasoning behind the award. Townes had proposed the idea of the microwave maser and later demonstrated it, earning him half of the prize. Basov and Prokhorov had published the first analytical description of the microwave maser, earning them the half of the prize that was split between them. Together they created the microwave maser, which in turn led to the laser.

A pure physicist in 1964 might think that "the maser-laser principle" was enough to earn a Nobel prize. That had been Goudsmit's attitude when he decreed *Physical Review Letters* need publish no more papers on masers. The principles were established, and he wasn't interested in the details. From our vantage point forty years later, that attitude seems myopic, mistaking the molehill of the maser for the mountain of the laser. Yet at least in the early 1960s, a pure physicist could legitimately argue that masers and lasers were fundamentally the same because they both relied on the amplification of stimulated emission. Townes and Bell Labs fuelled that attitude by merely adding the modifier "optical" to the label "maser." Looking back, Goudsmit's summary rejection of Maiman's paper might seem to merit an "Ig Nobel Prize." But his mistake was born of impatience and misunderstanding, not of stupidity.

If the prize was for the microwave maser, the Nobel Committee did not err badly in awarding it to Townes, Basov, and Prokhorov. A case could be made that Nicolaas Bloembergen deserved a share for inventing the three level solid state microwave maser, but the Nobel rules limit the prize to three people. In any case, he would share in the 1981 Nobel prize for his research on nonlinear optics and spectroscopy.

Yet in hindsight, the prize *should* have been awarded for the far more important optical laser. There could have been no denying Townes a share in that prize. He had carried the idea of stimulated emission beyond microwaves to the optical spectrum. He and Schawlow had outlined the principles of optical lasers just as Basov and Prokhorov had outlined those of microwave masers.

There should have been no denying Maiman a share in a laser prize, either. Maiman did not merely follow someone else's recipe. He carefully designed and executed a persuasive experiment that proved the principle. It was an experiment so solid that TRG, Bell, and others could duplicate it within weeks from newspaper reports. It used a material that Schawlow had said wouldn't work. It created a type of laser that Schawlow and Townes hadn't envisioned. It was Nobel quality work. In the words of Willis Lamb—a 1955 Nobel Laureate and Maiman's dissertation advisor—if Maiman had gotten the prize, "no one could have said with any reasonable justice that he shouldn't have gotten it."

Schawlow is a different matter. Although his optical insight played a key role in developing laser theory, ruby was an easy ground ball that dribbled between his legs. He proclaimed that pink ruby would never work, yet he had never studied it carefully enough to question the flawed data on ruby fluorescence efficiency. He thought red ruby might work, but never pursued it very seriously. Schawlow was everybody's friend, and his work on laser theory was important. But his heart was not in building devices, and it showed in his laser work. In 1981, he shared the Nobel Prize in Physics with Bloembergen for his later research using lasers for spectroscopy—a field where his heart belonged and his gifts shined brightly.

A few others deserve consideration.

Fabrikant proposed optical amplification by stimulated emission more than a decade before anyone else, although he didn't consider oscillation until others suggested it. His work links the earlier generation of spectroscopists with the later development of the laser. Yet his work was largely ignored at the time, and he did not follow through on it. That may not have been his fault, but it meant that he had little direct influence, and his achievements were hard to document. In addition, he and Butayeva were fooled by an instrumental artifact into reporting gain that didn't exist. It would have been nice to see the honor to go a solid scientist doing important work while veiled in obscurity by the Soviet system, but Fabrikant didn't quite fill the bill.

Basov and Prokhorov were the first to propose a three-level microwave maser, and Bloembergen came up with the same idea but went much further on his own. That energy scheme was a key step on the road to the optical laser, but it belonged more to the realm of microwave masers.

Javan and Bennett's helium–neon laser was an impressive achievement, the first gas laser and the first laser to generate a continuous beam. They excited the laser in a way Schawlow and Townes had not anticipated, although Gould did. It clearly would have made the two serious Nobel contenders—if Maiman hadn't made the ruby laser first.

If Gordon Gould's main patent application or ARPA proposal had circulated broadly, it would have been the basic road map for laser research. As a patent application it lacked the rigor of a scientific paper, but its scope was broader than the Schawlow–Townes paper. Gould had both the big picture and the little details, many of which would become important as laser technology developed. His ideas shared a common root with Townes's, but they evolved independently. Unfortunately, the security clearance mess, legal problems, competing claims, and fate conspired to obscure Gould's work for many years. The patents he received outside the United States in the 1960s received little attention, and his American claims were tied up in litigation for a couple of decades. The four U.S. patents he finally received covered only a fraction of his original ideas. Gould himself campaigned only for his patent claims, not for public recognition. Only in the 1980s did the importance of Gould's contributions become clear. Gould wouldn't have made the Nobel cut in 1964, but he would be a viable candidate today.

The Nobel Prize can be divided into at most three pieces, and Townes and Maiman would stand at the top of the list for a laser prize. Many other people made important contributions, but Gould stands the closest behind them.

To this day, Bell Labs lives in a state of denial, claiming it "invented" the laser based on the Schawlow–Townes optical maser paper. Yet it was Gould who eventually won the patent wars, despite a series of legal blunders. And none of the materials Schawlow and Townes initially proposed worked. Maiman's pulsed ruby design was the first laser to work, followed by Sorokin and Stevenson's rare earth, and Javan and Bennett's helium–neon.

The National Inventors Hall of Fame cites four people as inventors of the laser—Gordon Gould, Ted Maiman, Art Schawlow, and Charles Townes. That's a good list, but like any short list it is inevitably incomplete. Fabrikant, Bloembergen, Basov, and Prokhorov made important early steps. Sorokin and Stevenson showed ruby was no fluke and that four-level lasers were possible. Javan, Bennett, and Herriott overcame formidable problems to make the first continuous laser. Many other lasers have followed, as the technology evolved step by step over the decades since Gould walked away from his fateful conversation with Townes. The race to make the first laser was a vital part of that evolution. Ted Maiman won that race decisively. But there were other steps along the road and the technology will continue to evolve.

In the spring of 2000, Hughes Research Labs—now renamed simply HRL—had a party to celebrate the 40th anniversary of the laser. The parquet floor installed when Hughes bought the building had been removed, and the company handed out pieces as souvenirs. Maiman chose not to attend but invited many of the same people to his own party in Vancouver. I missed both celebrations but that summer stopped to tour the sprawling complex. My guide, Adrian Popa, showed me through the buildings and took me to the lab where Maiman, D'Haenens, and Asawa had worked, and the office across the hall where Maiman had analyzed his results.

By one of those strange twists of fate that put me on the trail of the laser in the first place, I found Maiman's old office occupied by an old Caltech classmate, Metin Mangir, who's spent most of his career working on lasers at Hughes. He conducts his laser experiments in the same lab where the laser was born.

When we first saw lasers in the 1960s, they were still young, curious glass tubes filled with helium and neon, emitting red beams in the electronic engineering lab. Our generation has seen lasers change from laboratory novelties to parts of everyday life. A laser prints the manuscript of this book to send to the publisher, a laser writes the CD-ROM I send along with it. I relax listening to a laser playing music from a CD. A semiconductor laser pointer sits among the pens in my desk drawer. It's easy to take these everyday lasers for granted. Yet when I aim the laser pointer at the wall and see its bright red spot sparkle, I think of the wonder of the first imperfect laser beam sparkling bright and new in the flash-dazzled eyes of Ted Maiman and Irnee D'Haenens.

DRAMATIS PERSONAE

Isaac Abella: graduate student of Townes at Columbia; tried to build a metal-vapor laser under Townes.

Paul Adams: patent attorney working at Advanced Research Projects Agency; ardent supporter of Gould's laser proposal.

Charles Asawa: assistant to Maiman at Hughes while a graduate student at University of California at Los Angeles; joined Hughes after earning doctorate.

Ed Ballik: Bell Labs technician on helium–neon laser project.

Nikolai Basov: microwave maser developer at Lebedev Physics Institute in Moscow; shared 1964 Nobel Prize with Prokhorov and Townes.

William Bennett: graduate student at Columbia until 1957, assistant professor at Yale until 1959, then worked with Javan on helium–neon laser at Bell Labs; later returned to Yale.

George Birnbaum: section head and later department head at Hughes; Ted Maiman's immediate supervisor.

Nicolaas Bloembergen: Dutch-born Harvard professor who proposed three-level solid-state microwave maser. Shared 1981 Nobel Prize in Physics with Art Schawlow.

Willard Boyle: Bell Labs laser researcher; later invented charge-coupled devices used in modern imaging systems.

Fatima Butayeva: worked with Fabrikant at Moscow Power Engineering Institute on optical amplification.

Albert Clogston: Bell Labs manager who coordinated laser projects; Art Schawlow's boss.

Robert C. Collins: worked with Don Nelson and Art Schawlow on first Bell Labs ruby laser.

Bill Culver: physicist at Rand Corporation and graduate student at University of California at Los Angeles; later cofounded Optelecom Inc. with Gordon Gould.

Herman Cummins: graduate student at Columbia under Townes, tried to build a metal vapor laser.

Malcolm Currie: one of three division managers at Hughes Research Laboratories; Lyons's boss.

Irnee D'Haenens: assistant to Maiman at Hughes; later earned doctorate in physics and remained at Hughes until retirement.

Richard Daly: manager of laser research project at TRG; later founder of laser company called Quantronix.

George Devlin: technician working for Art Schawlow at Bell Labs.

Robert Dicke: Princeton University physicist who proposed infrared version of maser.

Albert Einstein: proposed stimulated emission in 1916.

Viktor Evtuhov: physicist who worked with Maiman at Hughes.

Valentin Fabrikant: physics professor at Moscow Power Engineering Institute; first to propose optical amplification by stimulated emission.

Gardner Fox: developed first computer model of laser resonance between a pair of mirrors at Bell Labs with Tingye Li.

Peter Franken: graduate student at Columbia with Gould, professor at University of Michigan, later professor at University of Arizona.

Geoffrey Garrett: Bell Labs physicist who worked on calcium fluoride and early ruby lasers.

Stanley Geschwind: Bell Labs physicist who studied ruby properties.

Lawrence Goldmuntz: founder and president of TRG (Technical Research Group Inc.)

James Gordon: built first microwave maser as Ph.D. project for Townes at Columbia; later at Bell Labs.

Samuel Goudsmit: eminent Dutch-born physicist at the Brookhaven National Laboratory; editor in chief at the American Physical Society and founding editor of *Physical Review Letters*.

Gordon Gould: graduate student at Columbia, left to pursue laser patent; worked at TRG, later at Brooklyn College and Optelecom.

Andrew Haeff: director of Hughes Research Laboratories.

Oliver S. Heavens: British optics specialist who spent a 1959–1960 sabbatical at Columbia filling in for Townes.

Robert Hellwarth: physicist who worked with Maiman at Hughes; later professor at University of Southern California.

Donald R. Herriott: Bell Labs optics specialist who developed the optics for the helium-neon laser.

Ray Hoskins: physicist who worked on lasers with Maiman at Hughes and later at Korad.

Steve Jacobs: spectroscopist hired by TRG from Perkin-Elmer Corp. to work on potassium and cesium vapor lasers with Gould and Paul Rabinowitz.

Ali Javan: Iranian-born physicist who studied under Townes at Columbia until 1958, then joined Bell Labs where he developed the helium–neon laser with Bill Bennett and Don Herriott. Later at MIT.

Wolfgang Kaiser: Bell Labs physicist who worked on calcium fluoride and early ruby lasers.

Alfred Kastler: developed optical pumping at École Normale Supérieure in Paris for which he received 1966 Nobel Prize in Physics.

Robert Keegan: attorney who prepared Gould patent applications.

Chihiro Kikuchi: made first ruby microwave maser at University of Michigan Willow Run Laboratory in 1957.

Rudolf Kompfner: associate executive director for research at Bell Labs.

Polykarp Kusch: physics professor at Columbia, doctoral advisor to Gordon Gould; shared 1955 Nobel Prize in Physics with Willis Lamb.

Rudolf Ladenburg: German physicist who first observed stimulated emission.

Willis Lamb, Jr.: physics professor at Stanford from 1951 to 1956 where he was doctoral advisor to Ted Maiman and Irwin Wieder; later at Oxford University and University of Arizona; shared 1955 Nobel Prize in Physics with Polykarp Kusch for work done earlier at Columbia.

Benjamin Lax: semiconductor physicist at MIT Lincoln Laboratory.

Bela Lengyel: Hungarian-born physicist at Hughes; later at California State University at Northridge.

Harold Leventhal: Washington attorney who handled Gould's security clearance; later a Federal appeals judge.

Leo Levitt: physicist who shared a Hughes office with Charlie Asawa.

Tingye Li: developed first computer model of laser resonance between a pair of mirrors at Bell Labs with Gardner Fox.

Harold Lyons: invented cesium atomic clock at National Bureau of Standards in 1949; joined Hughes Research Laboratories in 1955 to head the atomic physics department; left Hughes in late 1960.

Theodore Maiman: graduate student at Stanford under Willis Lamb, Jr., then at Hughes Research Laboratories until 1961; left to start a laser group at Quantatron and later founded Korad.

Ron Martin: physicist worked on solid-state microwave masers at Bell Labs, then hired to develop solid-state lasers at TRG.

Bruce McAvoy: worked with Irwin Wieder at Westinghouse Research Labs.

Sid Millman: manager of physical research at Bell Labs.

Don Nelson: Bell Labs physicist who worked on early ruby lasers.

Maurice Newstein: theoretical physicist at TRG; later a professor at Brooklyn College and its successor, Polytechnic University.

Robert V. Pound: Harvard physicist; observed stimulated emission at 50 kilohertz in 1951.

Alexander Prokhorov: microwave maser developer at Lebedev Physics Institute in Moscow; shared 1964 Nobel Prize with Basov and Townes.

Edward M. Purcell: Harvard physicist, observed stimulated emission at 50 kilohertz in 1951.

I. I. Rabi: Columbia physicist and head of Columbia Radiation Laboratory; received 1944 Nobel Prize in Physics.

Paul Rabinowitz: experimental physicist hired by TRG; worked with Gould and Steve Jacobs on potassium and cesium vapor lasers.

John Sanders: Oxford University physicist who spent a sabbatical at Bell Labs on laser research from January to September 1959.

Arthur Schawlow: postdoctoral fellow under Townes at Columbia, then worked on superconductors and lasers at Bell Labs; later at Stanford University; shared Nobel Prize in Physics with Nicolaas Bloembergen in 1981.

Derrick Scovil: made first solid-state microwave maser at Bell Labs in late 1956.

Ben Senitzky: hired by TRG from Bell Labs to work on discharge-excited gas lasers; later at Brooklyn College.

George Smith: department manager at Hughes.

William V. Smith: managed microwave spectroscopy research at IBM; started Sorokin and Stevenson working on lasers.

Elias Snitzer: physicist at American Optical Research Laboratory who developed single-mode optical fibers, fiber lasers, and glass lasers.

Peter P. Sorokin: graduate student of Bloembergen at Harvard, developed second and third lasers at IBM Research Center with Mirek Stevenson.

Mirek Stevenson: graduate student of Townes at Columbia; developed second and third lasers at IBM with Peter P. Sorokin; later went into finance.

Malcolm Stitch: Hughes physicist.

Charles H. Townes: at Bell Labs until 1948; proposed microwave maser and optical laser concepts while physics professor at Columbia; took two-year leave to work at Institute for Defense Analyses 1959–1961; became provost at Massachusetts Institute of Technology; later moved to University of California at Berkeley.

Colin Webb: student of John Sanders at Oxford University.

Joseph Weber: electrical engineering professor at University of Maryland; proposed microwave amplification by stimulated emission.

Ken Wickersheim: Hughes physicist whose new spectrograph Lyons borrowed for early laser experiments.

Irwin Wieder: graduate student at Stanford under Willis Lamb, Jr., where he worked with Ted Maiman; studied masers, ruby, and optical pumping at Westinghouse Research Labs from 1956 to 1960, when he moved to Varian Associates.

Amnon Yariv: staff member at Bell Labs.

Adam Yarmolinsky: Washington attorney who handled Gould's security clearance; later an official in Kennedy, Johnson, and Carter administrations.

Herbert Zeiger: postdoctoral fellow under Townes at Columbia until 1953 working on microwave maser; later at MIT Lincoln Laboratory.

SOURCES

In writing this book I consulted a wide range of publications and documents, including original research papers, historical accounts and recollections, and transcripts of oral-history interviews recorded by other people. The bibliography that follows cites publications. I also have interviewed and corresponded with many people involved in the early development of lasers and microwave masers. Much of the correspondence has been electronic.

Three primary sources deserve special note because they give the viewpoints of three key players: Theodore Maiman, Gordon Gould, and Charles Townes. I highly recommend them if you want to learn more about the laser race, events leading up to it, and its aftermath.

Ted Maiman's autobiography, *The Laser Odyssey* (Laser Press, 2000), is an outspoken first-hand account by the man who made the laser happen. Read it to understand his viewpoint.

Nick Taylor's *Laser: The Inventor, the Nobel Laureate, and the Thirty-year Patent War* (Simon & Schuster, 2000) is an excellent account of Gordon Gould's long-running patent struggle, including a biography of Gould and details on TRG's laser program.

Charles Townes's autobiography, *How the Laser Happened: Adventures of a Scientist* (Oxford University Press, 1999), recounts a remarkably productive career that spans public service and astronomy as well as the microwave maser and the laser.

Three other books give historical accounts of laser development and have been useful resources for checking details.

Mario Bertolotti, *Masers and Lasers: An Historical Approach* (Adam Hilger Ltd., Bristol, UK, 1983). A physicist's account, heavy on early developments leading up to the maser and the laser.

Joan Bromberg, *The Laser in America 1950–1970* (MIT Press, 1991). A science historian's account of the research process, limited to the United States, written for the Laser History Project.

Jeff Hecht, *Laser Pioneers*, rev. ed. (Academic Press, Boston, 1991). A collection of interviews with early laser developers including Townes, Schawlow, Gould, Javan, Bloembergen, and Sorokin. It also includes a description of Maiman's work and an outline of laser history.

A few books collect important laser papers that are widely dispersed in the scientific literature:

Frank S. Barnes, editor, *Laser Theory* (IEEE Press, New York, 1972).

William T. Silfvast, editor, *Selected Papers on Fundamentals of Lasers* (SPIE Optical Engineering Press, Bellingham, Washington, 1993)

Jay R. Singer, editor, *Advances in Quantum Electronics* (Columbia University Press, New York, 1961). Proceedings of a major 1961 conference on masers and lasers.

Charles H. Townes, editor, *Quantum Electronics* (Columbia University Press, New York, 1960). Proceedings of the 1959 Shawanga Lodge conference.

HISTORICAL ACCOUNTS IN BOOKS AND JOURNALS

William R. Bennett Jr., "Background of an invention: the first gas laser," *IEEE Journal on Selected Topics in Quantum Electronics* 6, 869–875 (November/December 2000).

Nicolaas Bloembergen, *Encounters in Magnetic Resonances: Selected Papers of Nicolaas Bloembergen* (World Scientific, Singapore and New Jersey, 1996; see "Autobiographical notes," pp. 1–29).

Paul Forman, "Inventing the laser in postwar America," *Osiris 2nd series*, 7, 105–134 (1992).

Rufolf Kompfner, "The Invention of Traveling Wave Tubes," *IEEE Transactions on Electron Devices ED-23*, 730-738 (July 1976).

Bela Lengyel, "Evolution of masers and Lasers," *American Journal of Physics 34*, 903–313 (October 1966).

Arthur L. Schawlow, "Masers and Lasers," *IEEE Transactions on Electron Devices ED-23*, 773–779 (July 1976).

Arthur L. Schawlow, "From maser to laser," pp. 113–148 in Behram Kursunoglu and Arnold Perlmutter, *Impact of Basic Research on Technology* (Plenum, New York, 1973).

George F. Smith, "The early laser years at Hughes Aircraft Company," *IEEE Journal of Quantum Electronics QE-20*, 577–584 (June 1984).

Peter P. Sorokin, "Contributions of IBM toward the development of laser sources—1960 to the present," *IEEE Journal of Quantum Electronics QE-20*, 585–591 (June 1984).

ARCHIVAL SOURCES

The Center for the History of Physics at the American Institute of Physics has a rich collection of archives, including documents, correspondence, and oral-history interviews from the Laser History Project conducted during the 1980s. This collection has been invaluable, particularly in regard to people who were not available for further interviews. Material consulted includes interviews or documents from the following:

Isaac Abella
Nikolai Basov
Nicolaas Bloembergen
Fred Burns
Anthony DeMaria
Irnee D'Haenens
Robert Dicke
Walter Faust
Peter Franken
Lawrence Goldmuntz
Eugene Gordon
Gordon Gould
Robert Hellwarth
Donald Herriott
Rolf Landauer
Benjamin Lax
Theodore Maiman (Deposition on patent interference between Robert Hellwarth and Gordon Gould, dated 30 November 1967, supplied by George F. Smith to Laser History Project.)
Sidney Millman
Alexander Prokhorov
George Smith
Joseph Weber
Irwin Wieder
Amnon Yariv

I also consulted oral-history interviews with Charles Townes and Arthur Schawlow that the University of California at Berkeley has kindly made available on-line. (These are different from the interviews conducted by the Laser History Project.)

Arthur L. Schawlow. *Arthur L. Schawlow*. Regional Oral History Office, University of California, Berkeley, 1998. Available from the Online Archive of California, <http://ark.cdlib.org/ark:/13030/kt5b69n7k2>
Charles Hard Townes. *Charles Hard Townes*. Regional Oral History Office, University of California, Berkeley, 1994. Available from the Online Archive of California, <http://ark.cdlib.org/ark:/13030/kt3199n627>

I also obtained copies of documents from

Charles Asawa
Bell Labs
HRL (formerly Hughes Research Laboratories)

Bela Lengyel
National Institute of Standards and Technology (on Harold Lyons)
Stanford University (on Rudolf Kompfner)
Colin Webb
Irwin Wieder

Eugenia V. Fabrikant-Gamsakhourdia, L. M. Biberman, and Gennady Gorelik supplied English translations of articles by and about Valentin Fabrikant. Vladilen Letokhov supplied a tape recording of a lecture by Fabrikant, which Elina Zinkevich translated into English.

I have used some material from my own archives of interviews conducted for *Laser Pioneers* and during my research for *City of Light: The Story of Fiber Optics* (Oxford University Press, 1999). These are listed with the dates noted. All other interviews and correspondence listed were in connection with this book.

INTERVIEWS AND CORRESPONDENCE CONDUCTED BY AUTHOR

Isaac Abella (telephone)
Alexei Anikov (e-mail)
Marilyn Appel (e-mail)
Charlie Asawa (telephone, in person, written and e-mail correspondence)
Ed Ballik (telephone)
Dimitry Basov (telephone and e-mail)
Ben Bederson (telephone)
William Bennett (telephone and e-mail)
George Birnbaum (telephone and e-mail)
Martin Blum (telephone and e-mail)
Willard S. Boyle (telephone)
William Bridges (telephone)
Joan Lisa Bromberg (telephone)
Malcolm Butler (e-mail)
Geoffrey Charlish (e-mail forwarded by Malcolm Butler)
Bob Clauss (telephone)
Al Clogston (telephone and e-mail)
William Culver (e-mail and telephone)
Malcolm Currie (telephone)
Irnee D'Haenens (telephone and in person)
Bob Daly (e-mail)
George Devlin (e-mail)
Richard W. Dixon (telephone)
Viktor Evtuhov (telephone)
Paul Forman (telephone)
Ted Geballe (e-mail)
Joe Giordmaine (telephone)
Gennady Gorelik (in person, e-mail, telephone)
Gordon Gould (telephone and in person 1983 and 1984 for earlier book)

Art Guenther (telephone)
Robert Hellwarth (telephone for earlier book 1994)
Don Herriott (telephone)
Steve Jacobs (telephone and e-mail)
Steven Jarrett (telephone)
Ali Javan (in person 1985 and 1991 for earlier book)
Paul Kelley (telephone)
Robert H. Kingston (telephone)
Daniel Kleppner (telephone
Alexei Kojebnikov (e-mail)
Jack Kotik (e-mail)
Willis Lamb (telephone)
Benjamin Lax (telephone)
Bela Lengyel (telephone and e-mail)
Vladilen Letokhov (e-mail)
Tingye Li (telephone, in person, e-mail)
Roger Main (e-mail)
Ronald Martin (telephone)
Bruce McAvoy (telephone)
Mark McDermott (telephone)
H. Warren Moos (telephone)
Robert A Myers (telephone and e-mail)
Donald Nelson (in person and e-mail)
John Osmundsen (telephone)
Adrian Popa (telephone, in person, e-mail)
Paul Rabinowitz (telephone and e-mail)
Norman Ramsey (telephone)
Bronius Rinkevichius (e-mail)
Richard Samuel (telephone)
John Sanders (telephone)
Howard Schlossberg (telephone)
Ben Senitzky (telephone)
Tony Siegman (telephone and e-mail)
Eliot Sivowitch (telephone)
Peter Sorokin (in person 1985 for earlier book)
Elias Snitzer, (in person 1995 for earlier book)
Hermann Statz (telephone)
Boris Stoicheff (telephone)
Nick Taylor (telephone and e-mail)
Valerii V. Ter-Mikirtychev (telephone)
Charles Townes (in person)
Frances Townes (in person)
Colin Webb (telephone and e-mail)
Irwin Wieder (telephone, in person, written and e-mail correspondence)
Roger Woolnough (telephone)

Web sites have provided useful background information on the histories of companies including Hughes Research Labs and IBM. Unfortunately, these pages tend to move or disappear quickly, so there's little point in listing their addresses. The best way to find them is by using a search engine.

SOURCES FOR CHAPTERS

PREFACE

D'Haenens interview; Maiman, *Laser Odyssey*.

CHAPTER 1

Primary sources:
Bertolotti; Townes, *How the Laser Happened*.

Specific papers mentioned:
Albert Einstein, "The Quantum Theory of Radiation," reprinted in Frank S. Barnes, *Laser Theory* (IEEE Press, New York, 1972), 5–21.
V. A. Fabrikant, M. M. Vudynskii, and F. Butayeva, USSR patent No. 132209 submitted 18 June 1951, published 1959.
R. Ladenburg, *Rev Modern Physics 5*, 243 (1933).
E. M. Purcell and R. V. Pound, "A nuclear spin system at negative temperature," *Physical Review 81*, 279–280 (1951).
Richard C. Tolman, "Duration of molecules in upper quantum states," *Physical Review 23*, 693–709 (1924).

Other Sources:
Ali Javan suggested that stimulated emission went unrecognized; quoted in Jeff Hecht, *Laser Pioneers* (Academic Press, Boston, 1991) p. 163.
Information on Fabrikant collected from Eugenia V. Fabrikant-Gamsakhourdia, L. M. Biberman, Gennady Gorelik, recorded lecture by Fabrikant

CHAPTER 2

Primary sources:
Townes, *How the Laser Happened*; Bertolotti, Bromberg, Forman.

Papers cited:
N. G. Basov and A. M. Prokhorov, *Zh Eksp Teor Fiz*, 27, 431 (1954).
N. G. Basov and A. M. Prokhorov, "About possible methods for obtaining active molecules for a molecular oscillator," *Zh Eksp Teor Fiz 28* N. 2, 249–250 (February 1955); translated in *Soviet Physics JETP 1*, 184–185 (1956).

N. G. Basov and A. M. Prokhorov, "Theory of the molecular generator and molecular power amplifier," *Zh Eksp Teor Fiz*, 30(3), 560–564 (1956). [Carries a December 1954 submission date and references the short report published just after the maser worked, according to a translation by Morris D. Friedman. This paper also may have been given at the Faraday meeting.]

Nicolaas Bloembergen, "Proposal for a new type solid state maser," *Physical Review 104*, 324–327 (October 15, 1956).

J. P. Gordon, H. J. Zeiger, and C. H. Townes, "The maser—new type of microwave amplifier, frequency standard, and spectrometer," *Physical Review 99*, 1264–1274 (August 15, 1955).

J. P. Gordon, H. J. Zeiger, and C. H. Townes, "Molecular microwave oscillator and new hyperfine structure in the microwave spectrum of NH$_3$," *Physical Review 95*, 282–284 (July 1, 1954).

W. E. Lamb Jr. and R. C. Retherford, "Fine structure of the hydrogen atoms, Part I," *Physical Review 79*, 549.

E. M. Purcell and R. V. Pound, "A nuclear spin system at negative temperature," *Physical Review 82*, 279–280 (1951).

J. Weber, "Amplification of Microwave Radiation by Substances not in thermal equilibrium," *Transactions of the IRE Professional Group on Electron Devices*, PGED-3, June 1953, p. 1.

Other sources:

Basov, interview by Arthur Guenther for Laser History Project, 14 September 1984.

William Bennett, telephone interview, October 12, 2001.

Nicolaas Bloembergen, *Encounters in Magnetic Resonances: Selected Papers of Nicolaas Bloembergen* (World Scientific, Singapore and New Jersey, 1996; see "Autobiographical notes," pp. 1–29).

Bloembergen, interview with Joan Bromberg and Paul Kelley for Laser History Project.

James Gordon, interview with Paul Forman for Laser History Project.

Rudolf Kompfner, "The Invention of Traveling Wave Tubes," *IEEE Transactions on Electron Devices* ED-23, 730–738 (July 1976).

Alexander Prokhorov, interview for Laser History Project, 1984.

Townes and Schawlow, Berkeley oral history interviews.

CHAPTER 3

Primary Sources:

Townes, *How the Laser Happened*; Schawlow, "From maser to laser"; and Townes and Schawlow, Berkeley oral history interviews.

PUBLICATIONS:

Robert Dicke, "Molecular amplification and generation systems and methods," U.S. Patent 2,851,652, filed May 21, 1956, issued September 9, 1958.

A. M. Prokhorov, *Zh. Eksp. Teor Fiz 34*, 726 (1958).

A. L. Schawlow and C. H. Townes, "Masers and maser communications system," U.S. Patent 2,929,922, filed July 30, 1958, issued March 22, 1960.

A. L. Schawlow and C. H. Townes, "Infrared and optical masers," *Physical Review 112*, 1940–1949 (December 15, 1958).

Other sources:
William Bennett, e-mail to author June 30, 2003.
Bromberg; Taylor; Bertolotti; Townes, interview and oral history.
Gould, telephone interview, August 26, 2003.
Hecht, *Laser Pioneers*, Schawlow interview, p. 88.
Moos, telephone interview August 6, 2003.

CHAPTER 4:

Primary Sources:
Taylor; Gould interviews with author.

Publications:
A. L. Schawlow and C. H. Townes, "Infrared and optical masers," *Physical Review 112*, 1940–1949 (December 15, 1958).

Other Sources:
Marilyn Appel, e-mail, September 4, 2003.
Kotik, e-mail to author, dated March 18, 2003.
McDermott, telephone interview.
Rabinowitz, telephone interview.
Townes, *How the Laser Happened*

CHAPTER 5

Primary Sources:
Schawlow, oral history interview and "From maser to laser"; Townes, *How the Laser Happened*; Javan, interview with author; Bennett, "Background of an invention: the first gas laser"; Bromberg.

Scientific papers:
Arthur L. Schawlow, "Optical and Infrared Masers," pp. 553–562 in Charles H. Townes Ed, *Quantum Electronics* (Columbia University Press, New York, 1960).

Other Sources:
http://www.farhangsara.com/alijavan.html
Abella, paraphrase of interview for Laser History Project, 22 October 1985.
William Bennett, in e-mail supplied by Ted Geballe, 3 January 2002.
Clogston, telephone interview.
Herriott, "Laser Development Programs," addition to letter to Joan Bromberg, October 23, 1985.
Rudolf Kompfner, vita.

McDermott, telephone interview, Oct 22, 2001.

S. Millman, ed., *A History of Engineering and Science in the Bell System.*

Millman, interview with Joan Bromberg, April 25, 1986.

Sanders, telephone interview, December 4, 2001.

Orazio Svelto, *Principles of Lasers,* 2nd ed. (Plenum, New York, 1982).

CHAPTER 6

Primary Sources:
Taylor, Gould interviews

Other Sources:
Buderi, *The Invention That Changed the World.*
Cheney, *Tesla: Man out of Time.*
Baucom, *The Origins of SDI, 1944–1983.*
Bob Daly, interview.
Kotik e-mail.
Rabinowitz, interview.
Jacobs, telephone interview.
Culver, telephone interview.

CHAPTER 7

Primary Sources:
Maiman, *The Laser Odyssey*; Wieder, interview and correspondence.

Scientific papers and reports:
F. A. Butayeva and V. A. Fabrikant, "A medium with negative absorption," translated by Bela Lengyel (I have only the first 13 pages of typed MS). Originally published in *Investigations in Experimental and Theoretical Physics* (A memorial to G. S. Landsberg) (USSR Academy of Sciences Publications, Moscow, 1959), pp. 62–70.

J. E. Castle Jr. et al., "Quarterly Technical Report No. 5, Maser Studies," June 1958 to September 1958, Westinghouse Research Laboratories, dated Nov 17, 1958, copy supplied by I. Wieder.

Robert H. Dicke, U.S. Patent 2,851,652, "Molecular amplification and generation systems and methods," filed May 21, 1956, issued September 9, 1958.

A. M. Prokhorov, "Molecular amplifier and generator for submillimeter waves," *Soviet Physics JETP 7,* 1140–1141 (1958).

J. H. Sanders, M. J. Taylor, and C. E. Webb, "Search for Light Amplification in a mixture of mercury vapor and hydrogen," *Nature,* February 24, 1962, p. 767

Colin Webb, "Laser research at Oxford in the 1960s," *Optics and Photonics News 14* (5) 14–17 (May 2003).

Irwin Wieder, Exhibit 2, Disclosure to Western Electric Patent Department, dated March 21, 1958, showing date of conception January 15, 1958.

Irwin Wieder, Quarterly Technical Report No. 4, Maser Studies, Westinghouse Research Labs, Feb.–June 1958, pp. 1–4.

Other Sources:
Bromberg
Drosnin, *Citizen Hughes.*
Popa, interview.
National Bureau of Standards: "Biographical Sketch of Dr. Harold Lyons"
Jones, *Splitting the Second.*
George Smith, interview.
Lamb, "Laser theory and doppler effects."
D'Haenens, interview.
Lamb, interview.
George Smith, "The early laser years at Hughes Aircraft Company."
Martin, interview, Jan. 4, 2002.
Bloembergen interview, p. 103, in Hecht, *Laser Pioneers.*
Bennett, telephone interview Oct. 12, 2001.

CHAPTER 8

Primary Sources:
Townes, Oral History and *How the Laser Happened*; Schawlow, Oral History; Wieder, correspondence; Maiman, *Laser Odyssey*

Scientific papers:
Gordon Gould, "The Laser: Light amplification by stimulated emission of radiation," *Ann Arbor Conference on Optical Pumping*, June 15–18, 1959, p. 128–130.
Ali Javan, "Possibility of producing of negative temperature in gas discharge" *Physical Review Letters 3*, 87–89 (July 15, 1959).
Ali Javan, "Possibility of obtaining negative temperature in atoms by electron impact," pp. 564–571 in Charles H. Townes, ed., *Quantum Electronics* (Columbia University Press, New York, 1960).
J. H. Sanders, "Maser action in helium," *Ann Arbor Conference on Optical Pumping*, June 15–18, 1959, p. 131.
J. H. Sanders, "Optical maser design," *Physical Review Letters 3*, 86–87 (July 15, 1959).
Arthur Schawlow, "Discussion of Wieder's paper," *Ann Arbor Conference on Optical Pumping*, June 15–18, 1959, p. 137.
Arthur L. Schawlow, "Infrared and optical masers," pp. 553–562 in Charles H. Townes, ed., *Quantum Electronics* (Columbia University Press, New York, 1960).
Charles H. Townes, ed., *Quantum Electronics* (Columbia University Press, 1960).
Irwin Wieder, "A possible limitation on optical pumping in solids," *Ann Arbor Conference on Optical Pumping*, June 15–18, 1959, p. 133–136.
Irwin Wieder, "Solid-state, high-intensity monochromatic light sources," *The Review of Scientific Instrument 30*, 995–996 (November 1959).
Irwin Wieder, "Some microwave-optical experiments in ruby," pp. 105–109 in Charles H. Townes, ed., *Quantum Electronics* (Columbia University Press, New York, 1960).

Other Sources:
Abella, interviews

Bennett, "Background of an invention: the first gas laser."
Gould, interviews
Nelson, "Reminiscence of Schawlow."
Schawlow, "From maser to laser."
Siegman, correspondence
Sorokin, interview

CHAPTER 9

Primary sources:
Maiman, *Laser Odyssey*; Asawa interview and correspondence; D'Haenens interviews.

Scientific papers:
S. A. Goudsmit, "Editorial: Masers" *Physical Review Letters* 3, 125 (August 1, 1959).

Other sources:
Birnbaum, interview.
Bromberg
George Smith, interview.
Wieder, documents provided author

CHAPTER 10

Primary sources:
Schawlow, interview and "From maser to laser"; Bennett, interview and "Background of an invention"; Nelson, interview.

Scientific publications:
Willard S. Boyle and David G. Thomas, "Optical Maser," U.S. Patent 3,059,117, filed Jan 11, 1960, issued October 16, 1962.
C. H. Townes and A. L. Schawlow, *Microwave Spectroscopy* (McGraw-Hill, New York, 1955; reprinted by Dover Publications, New York, 1975).

Other sources:
Ballik, telephone interview, Feb. 27, 2002.
Clogston, telephone interview, January 7, 2002.
Devlin, interview and correspondence.
Faust, replies to questions for history of lasers, dated late May 1983, in files of Laser History Project.
Herriott, interview Oct. 12, 2001.
Li, interview and correspondence.
Millman, interview with Joan Bromberg, June 5, 1984.
Schawlow, "Masers and Lasers," *IEEE Transactions on Electron Devices* ED-23, 773–779 (July 1976).

CHAPTER 11

Primary sources:
Taylor; Gould interviews; Rabinowitz interview; Jacobs.

Scientific papers:

Gordon Gould, "Maser apparatus," British Patent Specification 953,721, U.S. application filed April 6, 1959; U.K. application filed April 1, 1960; published April 2, 1964.

Other sources:

Bob Daly, e-mail to author, 1 April 2003.
Goldmuntz, interview with Joan Bromberg.
Martin, interview, Jan 4, 2002.
Rabinowitz, interview.
Senitzky, interview, Jan 4, 2002.

CHAPTER 12

Primary sources:

Sorokin Interview, Snitzer interview.

Scientific publications:

Benjamin Lax, "Cyclotron resonances and impurity levels," in Charles Townes, ed., *Quantum Electronics* (Columbia University Press, New York, 1960), see ref. 40 on p. 447.
Irwin Wieder, "Optical detection of paramagnetic resonance saturation in ruby," *Physical Review Letters 3*, 468–470 (November 15, 1959).
Irwin Wieder, "Solid-state, high-intensity monochromatic light sources," *Review of Scientific Instruments 30*, 995–996 (November 1959).
Irwin Wieder, "Optically pumped maser and solid state light source for use therein," U.S. Patent 3,403,349, filed May 28, 1959, issued September 24, 1968.

Other sources:

Bennett, telephone interview Oct 12, 2001.
Bloom, "Optical pumping."
DeMaria, interview by John Bromberg, April 13, 1984.
Dupuis, "An introduction to the development of the semiconductor laser."
Jenkins and White, *Fundamentals of Optics.*
Landauer, interview for Laser History Project.
Lax, interview.
Sorokin, "Contributions of IBM."
Townes, oral history interview, Berkeley site.
Wieder, correspondence and interview.

CHAPTER 13

Primary sources:

Maiman, *Laser Odyssey*, Asawa, D'Haenens.

Scientific publications:

T. H. Maiman, "Optical and microwave experiments in ruby," *Physical Review Letters 4*, 564–566 (June 1, 1960).

Irwin Wieder, "Solid-state, high-intensity monochromatic light sources," *Review of Scientific Instruments* 30(11), 995–996 (November 1959).

Irwin Wieder, "Optical detection of paramagnetic resonance saturation in ruby," *Physical Review Letters* 3, 468–470 (November 15, 1959).

Other sources:

Birnbaum, telephone interview.

Edgerton, *Electronic Flash, Strobe.*

Evtuhov, telephone interview, June 15, 2000.

Lengyel, e-mail and interview.

Maiman, deposition on patent interference between Robert Hellwarth and Gordon Gould.

Popa, interview.

Schawlow, "From maser to laser."

George Smith, interview with Joan Bromberg for Laser History Project.

Smith, "Early laser years at Hughes Aircraft Company."

CHAPTER 14

Primary resources:

Abella, interviews; Jacobs, interviews; Rabinowitz; Taylor, Schawlow "From maser to laser" and oral history; Bennett interview and "Background of an invention."

Scientific publications:

F. A. Butayeva and V. A. Fabrikant, *Research in Experimental and Theoretical Physics,* Memorial Volume in Honor of G. S. Landsberg (USSR Academy of Sciences Press, Moscow, 1959) (Translation by Bela Lengyel; copy of pp. 1–13 supplied by Colin Webb. Other pages are missing.)

J. H. Sanders, M. J. Taylor, and C. E. Webb, "Search for light amplification in a mixture of mercury vapor and hydrogen," *Nature* February 24, 1962, p. 767.

Arthur Schawlow and Charles H. Townes, "Masers and maser communications systems," U.S. Patent 2,929,922, filed July 30, 1958, issued March 22, 1960.

Other sources:

Clogston, telephone interview, January 7, 2002.

Emma and Wolff, "Future Development in Engineering," *Electronics.*

Faust, "Reply to questions for history of lasers."

Geballe, e-mail dated January 1, 2002.

Li, e-mail and interview.

Martin, telephone interview, January 4, 2002.

Millman interview.

Nelson, interview, July 23, 2001.

Sanders, interview.

Senitzky, telephone interview, January 4, 2002.

Webb, telephone interview, June 18, 2001.

Wieder, telephone interview, June 12, 2001.

CHAPTER 15

Primary sources:
Maiman, *Laser Odyssey*; Irnee D'Haenens, interview; Asawa, interview and correspondence.

Scientific publications:
T. H. Maiman et al., "Stimulated optical emission in fluorescent solids II: Spectroscopy and stimulated emission in ruby," *Physical Review 123*, 1151–1157 (August 15, 1961).

Other sources:
Birnbaum, telephone interview, July 31, 2001.
Clogston, correspondence.
D'Haenens, Laser History Project document dated Aug 11, 1983, "History of maser research and development at Hughes Aircraft Company."
Hellwarth, telephone interview, July 25, 1994.
Jones, *Splitting the Second.*
Lengyel interview.
Maiman, in *Laser Pioneer Interviews.*
George Smith, interview.

CHAPTER 16

Primary sources:
Maiman, *Laser Odyssey*; D'Haenens, interview; Asawa, interview and correspondence.

Scientific papers:
T. H. Maiman, "Optical and microwave-optical experiments in ruby," *Physical Review Letters, 4*, 564–566 (June 1, 1960).
T. H. Maiman, *Journal of Applied Physics, 31*, p. 222 (1960).
T. H. Maiman, "Stimulated optical radiation in ruby," *Nature*, August 6, 1960, pp. 493–494.
T. H. Maiman, "TC1 Stimulated optical emission in ruby," *Journal of the Optical Society of America 50*, 1134 (November 1960).
T. H. Maiman, "Optical maser action in ruby," *British Journal of Communications and Electronics* Sept. 1960, 674–675.
T. H. Maiman, "Stimulated optical emission in fluorescent solids I: Theoretical considerations," *Physical Review 123*, 1145–1150 (August 15, 1961).
T. H. Maiman, R. H. Hoskins, I. J. D'Haenens, C. K. Asawa, and V. Evtuhov, "Stimulated optical emission in fluorescent solids II: Spectroscopy and stimulated emission in ruby," *Physical Review 123*, 1151–1157 (August 15, 1961).

Other sources:
"The death ray next," *Newsweek* July 18, 1960, pp. 78–79.
"Fantastic Red Spot" *Time* 76 pp. 47–48 (October 17, 1960).
"Light amplifier extends spectrum," *Electronics*, July 22, 1960, p. 43.
Abella, interview, August 6, 2001.

Bederson, telephone interview, August 30, 2001.

Bennett, telephone interview October 12, 2001.

Carl Byoir & Associates, press release for Hughes Aircraft, "U.S. victor in world quest of coherent light," dated July 7, 1960.

Charlish, e-mail forwarded by Malcolm Butler, dated 17 June 2003.

Dighton, "Death ray possibilities probed by scientists." Copy of undated newspaper clipping supplied by Adrian Popa. (Ads on reverse are from Pasadena, California.)

Franken, Laser History Project interview, March 8, 1985, p. 8.

Friedman, "Inventing the light fantastic: Ted Maiman and the world's first laser."

Lengyel, interview and correspondence.

Maiman, speech at a press conference at the Hotel Delmonico, New York, July 7, 1960. (Copy supplied by HRL Laboratories.)

Mann, "The month in science," *Popular Science,* 25–26 (October 1960).

Miller "Optical maser may aid space avionics," *Aviation Week 73,* 96–97 (July 18, 1960).

Moos, telephone interview, August 6, 2003.

Osmundsen, telephone interview, Feb 24, 2002.

Osmundsen, "Light amplification claimed by scientist," *New York Times,* July 8, 1960, pp. 1, 7.

Schawlow, "From maser to laser."

Snitzer, interview, June 12, 1995.

George Smith, Laser History Project interview, 5 February 1985.

Sullivan, "Air Force testing new light beam," *New York Times,* October 11, 1960, p. 16.

Townes, Berkeley oral history interview.

CHAPTER 17

Primary sources:

Maiman, *Laser Odyssey* and Deposition; Nelson interview; Schawlow "From maser to laser" and oral history interview.

Scientific publications:

R. J. Collins, D. F. Nelson, A. L. Schawlow, W. Bond, C. G. B. Garrett, and W. Kaiser, "Coherence, narrowing, directionality, and relaxation oscillations in the light emission from ruby," *Physical Review Letters 5,* 303–305 (October 1, 1960).

T. H. Maiman, "Optical maser action in ruby," *British Journal of Communications and Electronics.* Sept. 1960, pp. 674–676.

T. H. Maiman, "Stimulated optical emission in fluorescent solids I: Theoretical considerations," *Physical Review 123,* 1145–1150 (August 15, 1961).

T. H. Maiman, R. H. Hoskins, I. J. D'Haenens, C. K. Asawa, and V. Evtuhov, "Stimulated optical emission in fluorescent solids II: Spectroscopy and stimulated emission in ruby," *Physical Review 123,* 1151–1157 (August 15, 1961).

Theodore H. Maiman, paper TC1, "Stimulated Optical Emission in Ruby," *Journal of the Optical Society of America,* 50(11), 1125 (November 1960).

A. L. Schawlow, "Optical and infrared masers," pp. 553–563 in Charles Townes, ed., *Quantum Electronics* (Columbia University Press, New York, 1960).

A. L. Schawlow and G. E. Devlin, "Simultaneous optical maser action in two ruby satellite lines," *Physical Review Letters 6*, 96, (1961).

I. Wieder and L. R. Sarles, "Stimulated optical emission from exchange-coupled ions of Cr^{+++} in Al_2O_3," *Physical Review Letters 6*, 95 (1961).

Other sources:

"Fantastic red spot," *Time*, October 17, 1960, pp. 47–48.

"Optical maser used in communication experiments," press release dated October 5, 1960, from Bell Telephone Laboratories Inc., New York.

"Scientists demonstrate optical maser," *Electronics*, October 21, 1960, p. 38.

Abella, interviews.

Asawa, letter to author dated June 23, 2003.

Clogston, telephone interview, January 7, 2002.

Hecht, *City of Light: The Story of Fiber Optics*.

Jacobs, telephone interview, June 18, 2001.

Lamb, telephone interview June 16, 2001.

Martin, telephone interview, January 2, 2002.

Myers and Dixon, "Who Invented the Laser? An analysis of the Early Patents."

Sorokin interview.

Statz, telephone interview.

Boris Stoicheff, telephone interview, Oct 12, 2001.

Sullivan, "Air Force testing new light beam," *New York Times*, October 14, 1960, p. 16.

Townes, interview in Hecht, *Laser Pioneers*.

Yariv, "Catching the wave."

CHAPTER 18

Primary sources:

Sorokin interview, Javan interview; Bennett interview and "Background of an invention"; Herriott interview; Taylor.

Scientific publications:

V. K. Ablekov, M. S. Pesin and I. L. Fabelinskii, "The realization of a medium with negative absorption coefficient," *Zh Eksp Teor Fiz 39*, 892–893 (September 1960 in Russian); English Translation *Soviet Physics JETP 12*, 618–619 (March 1961).

W. R. Bennett Jr., "Gaseous optical masers," *Applied Optics Supplement on Optical Masers*, 24–61 (1962).

G. D. Boyd and J. P. Gordon, "Confocal multimode resonator for millimeter through optical wavelength masers," *Bell System Technical Journal 40*, 489–508 (1981).

Richard L. Daly, "Fluorescence of the Cr^{+++} pair spectrum in synthetic ruby," paper TB18, *Journal of the Optical Society of America 51* p. 473 (April 1961).

S. Jacobs, G. Gould, and P. Rabinowitz, "Coherent light amplification in optically pumped Cs vapor," *Physical Review Letters 7*, 415–417 (December 1, 1961).

Stephen Jacobs, Paul Rabinowitz, and Gordon Gould, "FA15 Optical pumping of cesium vapor" *Journal of the Optical Society of America 51*, 477 (April 1961).

Ali Javan, William R. Bennett Jr., and Donald R. Herriott, "Population inversion and continuous optical maser oscillation in a gas discharge containing a He-Ne mixture," *Physical Review Letters 6*, 106–110 (1961).

Ronald L. Martin, "Time development of the beam from the ruby laser," paper FA17 *Journal of the Optical Society of America 51*, 477 (April 1961).

P. Rabinowitz, S. Jacobs, and G. Gould, "Continuous optically pumped Cs laser," *Applied Optics 1*, 513–516 (July 1962).

P. P. Sorokin and M. J. Stevenson, "Stimulated infrared emission from trivalent uranium," *Physical Review Letters 5*, 557–559 (December 15, 1960).

Other Sources:

"Ookie the walrus gets X-ray of ailing tusk," *New York Times*, February 1, 1961, p. 39.

Ballik, telephone interview, Feb 27, 2002.

Bromberg, *The Laser in America*.

Jacobs, telephone interview, June 18, 2001.

Rabinowitz, telephone interview, March 18, 2003.

Schmeck, "Device outlined to amplify light," *New York Times*, December 15, 1960, p. 60.

Sullivan, "Bell shows beam of 'talking' light," *New York Times*, February 1, 1961, p. 39.

Witkin, "Space chimpanzee is safe after soaring 420 miles," *New York Times*, February 1, 1961, p. 1.

CHAPTER 19

Primary sources:

Maiman; Townes, *How the Laser Happened*; Schawlow interviews; Taylor, Gould interviews.

Scientific publications:

D. P. Devor, I. J. D'Haenens, and Charles K. Asawa, "Microwave generation in ruby due to population inversion produced by optical absorption," *Physical Review Letters 8*, 432–434 (June 1, 1962).

P. A. Franken, A. E. Hill, C. W. Peters, and G. Weinreich, "Generation of optical harmonics," *Physical Review Letters 7*, 118–119 (August 1, 1961).

R. N. Schwartz and Charles H. Townes, "Interstellar and interplanetary communication by optical masers," *Nature 190*, 205–208 (April 15, 1961).

Other sources:

"World's fair exhibit projects laser tree-clearing machine," *Laser Focus*, 14 (May 15, 1965).

Asawa, telephone interview, January 10, 2002.

D'Haenens, interview.

Ferreira-Marques, "Russian Nobel physics laureate Prokhorov dies," Reuters, Jan 8, 2002.

Gorelik, interview.

Gould, interview.

Lamb, telephone interview, June 16, 2001.

Lengyel, telephone interview, August 10, 2001.

Letokhov, e-mail to author.

Rinkevichius, e-mail, 13 March 2002.

Rinkevichius, "Distinguished Russian physicist celebrates 90th birthday" (*sic*), in "Briefly," *OE Reports 170* (February 1998).

Sagdeev, e-mail to author dated August 29, 2001.

BIBLIOGRAPHY

Ablekov, V. K., M. S. Pesin, and I. L. Fabelinskii, "The realization of a medium with negative absorption coefficient," *Zh Eksp Teor Fiz 39* 892–893 (September 1960 in Russian); English Translation Soviet Physics JETP 12 618 619 (March 1961).

Barnes, Frank S., editor, *Laser Theory (Selected reprints)* (IEEE Press, New York, 1972).

Bartusiak, Marcia, *Einstein's Unfinished Symphony* (Joseph Henry Press, Washington, 2000).

Basov, N. G. and A. M. Prokhorov, "Theory of the molecular generator and molecular power amplifier," *Zh Eksp Teor Fiz*, 30(3) 560 564 (1956). [Carries a December 1954 submission date and references the short report published just after the maser worked, according to a translation by Morris D. Friedman. This paper also may have been given at the Faraday meeting.]

Basov, N. G. and A. M. Prokhorov, *Zh Eksp Teor Fiz* 27, p. 431, (1954).

Basov, N. G. and A. M. Prokhorov, "About possible methods for obtaining active molecules for a molecular oscillator," *Zh Eksp Teor Fiz* 28 N. 2, 249–250 (February 1955); translated in *Soviet Physics JETP 1*, 184–185 (1956)

Baucom, Donald R., *The Origins of SDI, 1944–1983* (University Press of Kansas, Lawrence, 1992).

Bell Telephone Laboratories Inc., New York, press release dated October 5, 1960, "Optical maser used in communication experiments."

Bennett, William R. Jr., "Background of an invention: the first gas laser," *IEEE Journal on Selected Topics in Quantum Electronics 6*, 869–875 (Nov/Dec 2000).

Bennett, William R. Jr., "Gaseous optical masers," *Applied Optics Supplement on Optical Masers*, 24–61 (1962).

Bertolotti, Mario, *Masers and Lasers: An Historical Approach* (Adam Hilger Ltd., Bristol, 1983).

Bloembergen, Nicolaas, *Encounters in Magnetic Resonances: Selected Papers of Nicolaas Bloembergen* (World Scientific, Singapore and New Jersey, 1996); see "Autobiographical notes, pp. 1–29.

Bloembergen, Nicolaas, "Proposal for a new type solid state maser," *Physical Review 104*, 324–327 (October 15, 1956).

Bloom, Arnold L., "Optical pumping," *Scientific American 203* 4, 72–80 (October 1960).

Boyd, G. D. and J. P. Gordon, "Confocal multimode resonator for millimeter through optical wavelength masers," *Bell System Technical Journal 40*, 489–508 (1981).

Boyle, Willard S. and David G. Thomas, "Optical maser," U.S. Patent 3,059,117, filed January 11, 1960, issued October 16, 1962.

Bromberg, Joan Lisa, *The Laser in America 1950–1970* (MIT Press, Cambridge, 1991).

Buderi, Robert, *The Invention That Changed the World* (Simon & Schuster, New York, 1996).

Butayeva, F. A. and V. A. Fabrikant, "A medium with negative absorption," translated by Bela Lengyel, originally published in *Investigations in Experimental and Theoretical Physics* (A memorial to G. S. Landsberg) USSR Academy of Sciences Publications, Moscow, 1959, pp. 62–70 (copy of pp. 1–13 of translation supplied by Colin Webb, final pages missing).

Byoir, Carl & Associates, press release for Hughes Aircraft, "U.S. victor in world quest of coherent light," dated July 7, 1960.

Castle, J. E. Jr. et al., "Quarterly Technical Report No. 5, Maser Studies," June 1958 to September 1958, Westinghouse Research Laboratories, dated November 17, 1958, copy supplied by I. Wieder.

Cheney, Margaret, *Tesla: Man Out of Time* (Barnes & Noble Books, New York, 1993).

Collins, R. J., D. F. Nelson, A. L. Schawlow, W. Bond, C. G. B. Garrett, and W. Kaiser, "Coherence, narrowing, directionality, and relaxation oscillations in the light emission from ruby," *Physical Review Letters 5*, 303–305 (October 1, 1960).

Daly, Richard L., "Fluorescence of the Cr^{+++} pair spectrum in synthetic ruby," paper TB18, *Journal of the Optical Society of America 51*, 473 (April 1961).

"The Death Ray Next," *Newsweek*, July 18, 1960, pp. 78–79.

Devor, D. P., I. J. D'Haenens, and Charles K. Asawa, "Microwave generation in ruby due to population inversion produced by optical absorption," *Physical Review Letters 8*, 432–434 (June 1, 1962).

Dicke, Robert H., U.S. Patent 2,851,652, "Molecular amplification and generation systems and methods," filed May 21, 1956, issued September 9, 1958.

Dighton, Ralph, "Death ray possibilities probed by scientists." Copy of undated newspaper clipping supplied by Adrian Popa. (Ads on reverse are from Pasadena, California.)

Drosnin, Michael, *Citizen Hughes* (Holt, Rinehart & Winston, New York, 1985), p. 82.

Dupuis, Russell, "An introduction to the development of the semiconductor laser," *IEEE Journal of Quantum Electronics QE-20*, 651–657 (June 1984).

Edgerton, Harold E., *Electronic Flash, Strobe* (MIT Press, Cambridge, 1979).

Einstein, Albert, "The Quantum Theory of Radiation," reprinted in Frank S. Barnes, *Laser Theory* (IEEE Press, New York, 1972), pp. 5–21.

Emma, Thomas, and Michael F. Wolff, "Future Development in Engineering," *Electronics* March 4, 1960, 159–163. (See pp. 162–163.)

Fabrikant, V. A., *Selected Works* (Moscow, 2001, in Russian).

Fabrikant, V. A., M. M. Vudynskii, and F. Butayeva, USSR patent No. 132209 submitted 18 June 1951, published 1959.

"Fantastic Red Spot" *Time 76* (October 17, 1960) pp. 47–48.

Ferreira-Marques, Clara, "Russian Nobel physics laureate Prokhorov dies," Reuters, January 8, 2002.

Forman, Paul, "Inventing the laser in postwar America," *Osiris 2nd series*, 7, pp. 105–134 (1992).

P. A. Franken, A. E. Hill, C. W. Peters, and G. Weinreich, "Generation of optical harmonics," *Physical Review Letters 7*, 118–119 (August 1, 1961).

Friedman, Greg, "Inventing the light fantastic: Ted Maiman and the world's first laser," *OE Reports #200* August 2000 (no volume number) 5–6.

Garrett, C. G. B., *Gas Lasers* (McGraw-Hill, New York, 1967).

Gordon, J. P., H. J. Zeiger, and C. H. Townes, "The maser—new type of microwave amplifier, frequency standard, and spectrometer," *Physical Review 99*, 1264–1274 (August 15, 1955).

Gordon, J. P., H. J. Zeiger, and C. H. Townes, "Molecular microwave oscillator and new hyperfine structure in the microwave spectrum of NH_3," *Physical Review 95*, 282–284 (July 1, 1954).

Goudsmit, S. A., "Editorial: Masers" *Physical Review Letters 3*, 125 (August 1, 1959).

Gould, Gordon, "The laser: Light amplification by stimulated emission of radiation," *Ann Arbor Conference on Optical Pumping* (Conference Digest) June 15–18, 1959, pp. 128–130.

Gould, Gordon, "Maser apparatus," British Patent Specification 953,721, US Application filed April 6, 1959; UK application filed April 1, 1960; published April 2, 1964, p. 16.

Heavens, Oliver S., *Optical Masers* (Methuen & Co., London, 1964).

Hecht, Jeff, *City of Light: The Story of Fiber Optics* (Oxford University Press, New York, 1999).

Hecht, Jeff, *Laser Pioneer Interviews* (High Tech Publications, Torrance, CA, 1985).

Hecht, Jeff, *Laser Pioneers, Revised Edition* (Academic Press, Boston, 1992).

Jacobs, Stephen, G. Gould, and P. Rabinowitz, "Coherent light amplification in optically pumped Cs vapor," *Physical Review Letters 7*, 415–417 (December 1, 1961).

Jacobs, Stephen, Paul Rabinowitz, and Gordon Gould, "FA15 Optical pumping of cesium vapor" *Journal of the Optical Society of America 51*, 477 (April 1961).

Javan, Ali, "Possibility of obtaining negative temperature in atoms by electron impact," pp. 564–571 in Charles H. Townes, ed., *Quantum Electronics* (Columbia University Press, 1960).

Javan, Ali, "Possibility of producing of negative temperature in gas discharge" *Physical Review Letters 3*, 87–89 (July 15, 1959).

Javan, Ali, William R. Bennett Jr., and Donald R. Herriott, "Population inversion and continuous optical maser oscillation in a gas discharge containing a He-Ne mixture," *Physical Review Letters 6*, 106–110 (1961).

Jenkins, Francis A., and Harvey E. White, *Fundamentals of Optics*, 2nd ed. (McGraw-Hill, New York, 1950).

Jones, Tony, *Splitting the Second: The Story of Atomic Time* (Institute of Physics Publishing, Bristol and Philadelphia, 2000).

Kopfner, Rudolf, "The invention of traveling wave tubes," *IEEE Transactions on Electron Devices ED-23*, 730–738 (July 1976).

Ladenburg, Rudolf, "Dispersion in electrically excited gases," *Reviews of Modern Physics 5*, 243 (1933).

Lamb, Willis E. Jr., "Laser theory and doppler effects," *IEEE Journal of Quantum Electronics QE-20*, 551–555 (June 1984).

Lamb, Willis E. Jr. "Physical concepts in the development of the maser and laser," pp. 59–111 in Behram Kursunoglu and Arnold Perlmutter, eds., *Impact of Basic Research on Technology* (Plenum, New York, 1973).

Lamb, Willis E. Jr. and R. C. Retherford, "Fine structure of the hydrogen atoms, Part I," *Physical Review 79*, 549.

Lax, Benjamin, "Cyclotron resonances and Impurity Levels," in Charles Townes, ed., *Quantum Electronics* (Columbia University Press, 1960), see ref. 40 on p. 447.

Lengyel, Bela, A., "Evolution of masers and lasers," *American Journal of Physics 34*, 903–313 (October 1966).

Lengyel, Bela A., *Lasers: Generation of Light by Stimulated Emission* (Wiley, New York, 1962).

Levine, Albert K., ed., *Lasers Vol. 1* (Marcel Dekker, New York, 1966).

"Light amplifier extends spectrum," *Electronics, 43* (July 22, 1960).

Maiman, Theodore H., *The Laser Odyssey* (Laser Press, Blaine, Washington, 2000).

Maiman, T. H., "Maser behavior: temperature and concentration effects," *Journal of Applied Physics 31*, 222–223 (January 1960).

Maiman, T. H., "Optical and microwave-optical experiments in ruby," *Physical Review Letters 4*, 564–566 (June 1, 1960).

Maiman, T. H., "Stimulated optical radiation in ruby," *Nature 187*, 493–494 (August 6, 1960).

Maiman, T. H., "Optical maser action in ruby," *British Communications & Electronics, 674–676* (September 1960).

Maiman, T. H., "TC1 Stimulated optical emission in ruby," *Journal of the Optical Society of America 50*(11), 1134 (November 1960).

Maiman, T. H., "Stimulated optical emission in fluorescent solids I: Theoretical considerations," *Physical Review 123*, 1145–1150 (August 15, 1961).

Maiman, T. H., R. H. Hoskins, I. J. D'Haenens, C. K. Asawa, and V. Evtuhov, "Stimulated optical emission in fluorescent solids II: Spectroscopy and stimulated emission in ruby," *Physical Review 123*, 1151–1157 (August 15, 1961).

Maiman, Theodore H., speech at a press conference at the Hotel Delmonico, New York, July 7, 1960 (copy supplied by HRL Laboratories).

Maiman, T. H., "Ruby laser systems" U.S. Patent 3,353,115, filed November 29, 1965, issued November 14, 1967.

Mann, Martin, "The month in science," *Popular Science*, 25–26 (October 1960).

Martin, Ronald L., "Time development of the beam from the ruby laser," paper FA17 *Journal of the Optical Society of America 51*, 477 (April 1961).

Miller, B., "Optical maser may aid space avionics," *Aviation Week 73* July 18, 1960, pp. 96–97.

Millman, Sidney, ed., *A History of Engineering & Science in the Bell System: Physical Sciences (1925–1980)* (AT&T Bell Laboratories, no city 1983).

Myers, Robert A., and Richard W. Dixon, "Who invented the laser? An analysis of the Early Patents," unpublished, supplied by authors.

National Bureau of Standards, "Biographical sketch of Dr. Harold Lyons," National Bureau of Standards, Washington, undated (Faxed by NIST archivist March 14, 2002).

Nelson, Donald F., "Reminiscence of Schawlow at the first conference on lasers," pp. 121–122 in W. M. Yen and M. D. Levenson, eds., *Lasers, Spectroscopy, and New Ideas: A Tribute to Arthur L. Schawlow* (Springer-Verlag, New York, 1987).

"Ookie the walrus gets x-ray of ailing tusk," *New York Times*, February 1, 1961, p. 39.

Osmundsen, John A., "Light amplification claimed by scientist," *New York Times*, July 8, 1960, pp. 1, 7.

Patel, C. K. N., "Lasers—their development and applications at AT&T Bell Laboratories," *IEEE Journal of Quantum Electronics QE-20*, 561–576 (June 1984).

Prokhorov, A. M., "Molecular amplifier and generator for submillimeter waves," *Soviet Physics JETP 7*, 1140–1141 (1958)

Prokhorov, A. M. *Zh Eksp Teor Fiz 34* (1958) 762.

Purcell, E. M. and R. V. Pound, "A nuclear spin system at negative temperature," *Physical Review 82*, 279–280 (1951).

Rabinowitz, Paul., S. Jacobs, and G. Gould, "Continuous optically pumped Cs laser," *Applied Optics 1*, 513–516 (July 1962).

Rinkevichius, Bronius, "Distinguished Russian physicist celebrates 90th birthday," in "Briefly," *OE Reports 170* (February 1998).

Sanders, J. H., "Maser action in helium," *Ann Arbor Conference on Optical Pumping* (Conference Digest), June 15–18, 1959, p. 131.

Sanders, J. H., "Optical maser design," *Physical Review Letters 3*, 86–87 (July 15, 1959).

Sanders, J. H., M. J. Taylor, and C. E. Webb, "Search for light amplification in a mixture of mercury vapor and hydrogen," *Nature 113*, 767 (Feb 24, 1962).

Schawlow, Arthur L., "Masers and lasers," *IEEE Transactions on Electron Devices ED 23*, 773–779 (July 1976).

Schawlow, Arthur L., "Optical and infrared masers," pp. 553–563 in Charles Townes, ed., *Quantum Electronics* (Columbia University Press, New York, 1960).

Schawlow, Arthur L., "From maser to laser," pp. 113–148 in Behram Kursunoglu and Arnold Perlmutter, ed., *Impact of Basic Research on Technology* (Plenum, New York, 1973).

Schawlow, Arthur L., "Lasers in historical perspective," *IEEE Journal of Quantum Electronics QE-20*, 558–561 (June 1984).

Schawlow, Arthur, "Discussion of Wieder's paper," *Ann Arbor Conference on Optical Pumping*, June 15–18, 1959, p. 137.

Schawlow, A. L. and G. E. Devlin, "Simultaneous optical maser action in two ruby satellite lines," *Physical Review Letters 6*, 96, 1961.

Schawlow, A. L. and C. H. Townes, "Infrared and optical masers," *Physical Review 112*, 1940–1949 (December 15, 1958).

Schawlow, A. L. and C. H. Townes, "Masers and maser communications system," U.S. Patent 2,929,922, filed July 30, 1958, issued March 22, 1960.

Schmeck, Harold M. Jr., "Device outlined to amplify light," *New York Times*, December 15, 1960, p. 60.

Schwartz, R. N. and Charles H. Townes, "Interstellar and interplanetary communication by optical masers," *Nature 190*, 205–208, April 15, 1961.

"Scientists demonstrate optical maser," *Electronics, 38* (October 21, 1960).

Siegman, Anthony E., *Microwave and Solid-state Masers* (McGraw-Hill, New York, 1964).

Siegman, Anthony E., "Laser beams and resonators: the 1960s" *IEEE Journal on Selected Topics in Quantum Electronics 6*, 1380–1388 (November/December 2000).

Silfvast, William T., ed., *Selected Papers on Fundamentals of Lasers* (SPIE Optical Engineering Press, Bellingham, Washington, 1993).

Singer, Jay R., ed., *Advances in Quantum Electronics* (Columbia University Press, New York, 1961).

Smith, George F. "The early laser years at Hughes Aircraft Company," *IEEE Journal of Quantum Electronics QE-20*, 577–584 (June 1984).

Sorokin, Peter P., "Contributions of IBM toward the development of laser sources—1960 to the present," *IEEE Journal of Quantum Electronics QE-20*, 585–591 (June 1984).

Sorokin, P. P. and M. J. Stevenson, "Stimulated infrared emission from trivalent uranium," *Physical Review Letters 5*, 557–559 (December 15, 1960).

Sullivan, Walter, "Air Force testing new light beam," *New York Times*, October 14, 1960, p. 16.

Sullivan, Walter, "Bell shows beam of 'talking' light," *New York Times*, February 1, 1961, p. 39.

Svelto, Orazio, *Principles of Lasers*, 2nd ed. (Plenum, New York, 1982).

Tarasov, L. V., *Laser Age in Optics* (Mir Publishers, Moscow, 1981).

Taylor, Nick, *Laser: The Inventor, The Nobel Laureate, and the Thirty-Year Patent War* (Simon & Schuster, New York, 2000).

Tolman, Richard C., "Duration of molecules in upper quantum states," *Physical Review 23*, 693–709 (1924).

Townes, Charles H., ed. *Quantum Electronics* (Columbia University Press, New York, 1960).

Townes, Charles H., *How the Laser Happened: Adventures of a Scientist* (Oxford University Press, New York, 1999).

Townes, C. H. and A. L. Schawlow, *Microwave Spectroscopy* (McGraw-Hill, New York, 1955; reprinted by Dover Publications, New York, 1975).

Webb, Colin, "Laser research at Oxford in the 1960s," *Optics and Photonics News 14*(5), 14–17 (May 2003).

Weber, Joseph, "Amplification of microwave radiation by substances not in thermal equilibrium," *Transactions of the IRE Professional Group on Electron Devices, PGED-3*, 1 (June 1953).

Wieder, Irwin, "A possible limitation on optical pumping in solids," *Ann Arbor Conference on Optical Pumping* (Conference Digest) June 15–18, 1959, pp. 133–136.

Wieder, Irwin, "Solid-state, high-intensity monochromatic light sources," *Review of Scientific Instruments 30*(11), 995–996 (November 1959).

Wieder, Irwin, "Optical detection of paramagnetic resonance saturation in ruby," *Physical Review Letters 3*, 468–470 (November 15, 1959).

Wieder, Irwin, "Optically pumped maser and solid state light source for use therein," U.S. Patent 3,403,349, filed May 28, 1959, issued September 24, 1968.

Wieder, Irwin, "Some microwave-optical experiments in ruby," pp. 105–109 in Charles H. Townes, ed., *Quantum Electronics* (Columbia University Press, 1960).

Wieder, Irwin, *Quarterly Technical Report No. 4, Maser Studies*, Westinghouse Research Labs, Feb.–June 1958, pp. 1–4.

Wieder, Irwin, and L. R. Sarles, "Stimulated optical emission from exchange-coupled ions of Cr^{+++} in Al_2O_3," *Physical Review Letters 6*, 95 (1961).

Witkin, Richard, "Space chimpanzee is safe after soaring 420 miles," *New York Times*, February 1, 1961, p. 1.

"World's fair exhibit projects laser tree-clearing machine," *Laser Focus*, 14 (May 15, 1965).

Yariv, Amnon, "Catching the wave," *IEEE Journal on Selected Topics in Quantum Electronics 6*, 1478–1489 (November/December 2000).

INDEX

Italic page numbers indicate figures.